Advanced Technology-Assisted Problem Solving in Engineering Education:

Emerging Research and Opportunities

Manjit Singh Sidhu
Universiti Tenaga Nasional (UNITEN), Malaysia

A volume in the Advances in Educational Technologies and Instructional Design (AETID) Book Series

IGI Global
DISSEMINATOR OF KNOWLEDGE

Published in the United States of America by
IGI Global
Engineering Science Reference (an imprint of IGI Global)
701 E. Chocolate Avenue
Hershey PA, USA 17033
Tel: 717-533-8845
Fax: 717-533-8661
E-mail: cust@igi-global.com
Web site: http://www.igi-global.com

Library of Congress Cataloging-in-Publication Data

Names: Sidhu, Manjit Singh, 1971- author.
Title: Advanced technology-assisted problem solving in engineering education : emerging research and opportunities / Manjit Singh Sidhu.
Description: Hershey, PA : Engineering Science Reference, 2019. | Includes bibliographical references and index. | Summary: "This book examines issues and trends in computer-aided learning in engineering education"-- Provided by publisher.
Identifiers: LCCN 2019023058 (print) | LCCN 2019023059 (ebook) | ISBN 9781799804659 (hardcover) | ISBN 9781799804666 (paperback) | ISBN 9781799804673 (ebook)
Subjects: LCSH: Engineering--Study and teaching. | Computer-assisted instruction.
Classification: LCC T65.5.C65 .S558 2019 (print) | LCC T65.5.C65 (ebook) | DDC 620.0078/5--dc23
LC record available at https://lccn.loc.gov/2019023058
LC ebook record available at https://lccn.loc.gov/2019023059

This book is published in the IGI Global book series Advances in Educational Technologies and Instructional Design (AETID) (ISSN: 2326-8905; eISSN: 2326-8913)

British Cataloguing in Publication Data
A Cataloguing in Publication record for this book is available from the British Library.

All work contributed to this book is new, previously-unpublished material.
The views expressed in this book are those of the authors, but not necessarily of the publisher.

For electronic access to this publication, please contact: eresources@igi-global.com.

Advances in Educational Technologies and Instructional Design (AETID) Book Series

ISSN:2326-8905
EISSN:2326-8913

Editor-in-Chief: Lawrence A. Tome, Robert Morris University, USA

MISSION

Education has undergone, and continues to undergo, immense changes in the way it is enacted and distributed to both child and adult learners. In modern education, the traditional classroom learning experience has evolved to include technological resources and to provide online classroom opportunities to students of all ages regardless of their geographical locations. From distance education, Massive-Open-Online-Courses (MOOCs), and electronic tablets in the classroom, technology is now an integral part of learning and is also affecting the way educators communicate information to students.

The **Advances in Educational Technologies & Instructional Design (AETID) Book Series** explores new research and theories for facilitating learning and improving educational performance utilizing technological processes and resources. The series examines technologies that can be integrated into K-12 classrooms to improve skills and learning abilities in all subjects including STEM education and language learning. Additionally, it studies the emergence of fully online classrooms for young and adult learners alike, and the communication and accountability challenges that can arise. Trending topics that are covered include adaptive learning, game-based learning, virtual school environments, and social media effects. School administrators, educators, academicians, researchers, and students will find this series to be an excellent resource for the effective design and implementation of learning technologies in their classes.

COVERAGE

- Higher Education Technologies
- Social Media Effects on Education
- Bring-Your-Own-Device
- Web 2.0 and Education
- Adaptive Learning
- Digital Divide in Education
- Educational Telecommunications
- Hybrid Learning
- Classroom Response Systems
- Collaboration Tools

IGI Global is currently accepting manuscripts for publication within this series. To submit a proposal for a volume in this series, please contact our Acquisition Editors at Acquisitions@igi-global.com or visit: http://www.igi-global.com/publish/.

The Advances in Educational Technologies and Instructional Design (AETID) Book Series (ISSN 2326-8905) is published by IGI Global, 701 E. Chocolate Avenue, Hershey, PA 17033-1240, USA, www.igi-global.com. This series is composed of titles available for purchase individually; each title is edited to be contextually exclusive from any other title within the series. For pricing and ordering information please visit http://www.igi-global.com/book-series/advances-educational-technologies-instructional-design/73678. Postmaster: Send all address changes to above address.

Titles in this Series

For a list of additional titles in this series, please visit:
https://www.igi-global.com/book-series/advances-educational-technologies-instructional-design/73678/

Revolutionizing Education in the Age of AI and Machine Learning
Maki K. Habib (The American University in Cairo, Egypt)
Information Science Reference • copyright 2020 • 264pp • H/C (ISBN: 9781522577935) • US $185.00 (our price)

Form, Function, and Style in Instructional Design Emerging Research and Opportunities
Shalin Hai-Jew (Kansas State University, USA)
Information Science Reference • copyright 2020 • 203pp • H/C (ISBN: 9781522598336) • US $155.00 (our price)

The Roles of Technology and Globalization in Educational Transformation
Blessing F. Adeoye (Walden University, USA) and Gladys Arome (Marian University, USA)
Information Science Reference • copyright 2020 • 259pp • H/C (ISBN: 9781522597469) • US $195.00 (our price)

Utilizing Educational Data Mining Techniques for Improved Learning Emerging Research and Opportunities
Chintan Bhatt (Charotar University of Science and Technology, India) Priti Srinivas Sajja (Sardar Patel University, India) and Sidath Liyanage (University of Kelaniya, Sri Lanka)
Information Science Reference • copyright 2020 • 166pp • H/C (ISBN: 9781799800101) • US $165.00 (our price)

Claiming Identity Through Redefined Teaching in Construction Programs
Sherif Mostafa (Griffith University, Australia) and Payam Rahnamayiezekavat (Western Sydney University, Australia)
Information Science Reference • copyright 2020 • 259pp • H/C (ISBN: 9781522584520) • US $185.00 (our price)

For an entire list of titles in this series, please visit:
https://www.igi-global.com/book-series/advances-educational-technologies-instructional-design/73678/

IGI Global
DISSEMINATOR OF KNOWLEDGE

701 East Chocolate Avenue, Hershey, PA 17033, USA
Tel: 717-533-8845 x100 • Fax: 717-533-8661
E-Mail: cust@igi-global.com • www.igi-global.com

This book is dedicated to my wife, Dr. Kirandeep Kaur Sidhu, for her patience and forbearance, my daughter, Tavleen Kaur Sidhu, and my late mother, Bachan Kaur Sidhu.

Table of Contents

Preface

Global connectivity, smart machines, and new visual media are just some of the drivers reshaping how we think about education, what constitutes education, and how we learn and develop the skills to work in the future. Educationalists debate the many ways in which the content of education – at all levels – and the process of learning, will need to change over the years ahead. Whether it is classroom or workplace, online or offline, structured or unstructured, taught or learnt, standardized or not, certificated or not, then learning is likely to break free from our old mindsets in the coming years. In responds to the needs of "industry 4.0" or the fourth industrial revolution, where man and machine align to enable new possibilities, many academicians are altering their teaching techniques today. One well known media that is being used extensively today is visual multimedia which also includes mix or hybrid realities. Visual multimedia is captivating academicians because of its strengths to communicate difficult concepts in simple yet efficient ways. The key area lies in the learning environment. It is unstructured, and the learner is free to navigate the vast universe of information. Visual multimedia applications integrate animation, sound, graphics, and video to create an engaging, interactive and effective learning environment. Such software allows students to exercise more control over the pacing and sequencing of their own learning. With the availability of more sophisticated computers the potential to employ multimedia has grown tremendously.

In the case of engineering education which depends on static images, diagrams and complex mathematical computations in the traditional classroom, the employment of visual multimedia in teaching provides a better treatment in enriching the learning experiences by providing a multi-sensory perspective. In general, while it has been acknowledged that visual multimedia can be useful in the teaching of engineering subjects, non information technology academicians tend to believe that development of multimedia applications requires wide knowledge of high-end hardware, software and programming skills. This is true only partially. Today, with the availability of multimedia authoring tools such as Adobe Captivate®, Adobe Director®, Adobe Animate CC®, Authorware®, click2learn ToolBook® and 3D modeling, animation and rendering tools such as Alias Maya®, Autodesk 3D Studio Max®, Maxon Cinema 4D®, NewTek Lightwave® etc, we can develop multimedia materials with a little practice of these tools.

However, knowledge of the above mentioned authoring and modeling tools is not enough to develop a high quality teaching package that students could use in their learning. In order to be useful and transfer adequate knowledge skills to the learner, a multimedia programme design needs to have a sound pedagogical base. This book intends to help academicians with particular in the mechanical engineering field (although would be beneficial for other domains of engineering and sciences) in understanding the basic concepts of multimedia and various issues involved in the development of advanced multimedia problem solving packages.

The main purpose of this book is to share knowledge of issues and trends in computer aided learning (CAL) in particular, for engineering, from the perspectives of problems faced by students and instructors, students preference of learning styles and a new approach to learning, visualizing and problem solving. In addition this book has presented some interesting challenges in the development of new problem solving packages that uses the principles of CAL. These integrated packages are termed as Advanced Technology Assisted Problem Solving or ATAPS packages, which guide students step-by-step to complete various engineering mechanics problems.

As a use case the outcomes of the research presented in this book focused on a higher learning institution in Malaysia. Since the use of CAL in higher learning institutions in Malaysia is gaining wide attention, this study is mainly concerned with the development of effective ATAPS packages in supplementing the teaching and learning of engineering mechanics.

The study adopted multi-design approach (i.e., 2-D, 3-D, coach-based, and desktop virtual and augmented realities) to simplify the underlying engineering principles and thereby accelerate the learning process of slow learners (i.e. learners experiencing difficulties with understanding Engineering Mechanics theories). Several ATAPS packages were developed and tested by undergraduates to validate the design approach of the ATAPS packages. These ATAPS packages were developed using various 3-D modeling and multimedia authoring tools. The ATAPS packages were structured according to the learners' needs based on the survey carried out using the Felder-Solomon's ILS questionnaires and the packages were evaluated using various techniques for its effectiveness. Four groups of learners were identified i.e. sensory, visual, active and sequential. The results showed that different group of learners have different preferences of the features offered in the ATAPS packages. Nevertheless, in most of the experimental work, it was found that the step-by-step approach which was integrated in all the ATAPS packages was beneficial in promoting learning and understanding of Engineering Mechanics concepts, particularly to slow learners. The outcome of this study indicates that the ATAPS packages have great potential in aiding the learning of engineering and to enhance students' visualization in solving Engineering Mechanics problems.

ORGANIZATION OF THE BOOK

In Chapter 1, the author presents the problems faced by educators and students in teaching and learning engineering courses. In general teaching of engineering courses is relatively challenging due to the nature of the courses that are perceived as "difficult courses" by a number of students. This course is built upon the strong fundamental knowledge in physics and mathematics. This course requires the students to have a strong abstract thinking, reasoning and problem solving skills. Due to this problem a further research study (a continuation from previous study) was embarked at the National Energy University which is also known as "Universiti Tenaga Nasional, (UNITEN)" Malaysia. Selected lecturers and students were randomly selected and interviewed to find out the reasons to this problem. Thus, the purpose of this research was to explore the problems faced by mechanical engineering (ME) lecturers teaching the course and students studying difficult topics i.e. four link bar (4BL) mechanisms, shaft crank, motions such as linear, projectile and curvilinear, spur gear design and rotational of x and y axes. In response to the new paradigm shift in engineering education that emphasizes on the use of Information and Communication Technologies (ICT) to facilitate the teaching and learning for tertiary education, there have been numerous studies that proposed the solutions utilizing the Computer Aided Learning (CAL) platform. The usability of the CAL software to facilitate the learning of the students need further investigation empirically. In this research, several advanced technology assisted problem solving (ATAPS) packages which can be considered as a branch of CAL was enhanced with simulations and augmented reality (AR) technologies. Additionally, this research was also aimed at identifying the students' attitude towards outcome based education

(OBE). Literature has shown that OBE could enhance students' understanding and make the learning more meaningful. The theoretical framework of this study was based on selected learning styles instruments and software measurement inventory such as Honey and Mumford, Felder and Silverman, the Ogden's Personality and Learning Styles Questionnaire. The research study further investigated the usability of the enhanced ATAPS packages. Through the usability testing, two of the usability evaluation instruments namely (i) system usability scale (SUS) and (ii) post-study system usability questionnaire (PSSUQ) were employed which will be further discussed in other Chapters of this book.

Chapter 2 reviewed the literature on the trends and paradigm shifts in engineering education. Through the literature, the problems in mechanical engineering specifically for the learning of mechanics dynamics were identified. The importance of Information and Communication Technologies (ICT) and the application of computer aided learning (CAL) in engineering education was discussed. In the last section of this Chapter, the theory of learning styles associated with engineering education was studied.

In Chapter 3, the user interface design and the grounded learning theories are discussed. User interface (UI) design is the process of making interfaces in software or computerized devices with a focus on looks or style. Designers aim to create designs users will find easy to use and pleasurable. IU design typically refers to graphical user interfaces but also includes others, such as voice-controlled ones. Next, the interaction styles and the types of interactions are discussed. The usability benchmark and the usability evaluation instruments are discussed also in this chapter.

Chapter 4, the synthesis and analysis of four link bar (4BL) mechanisms problem which is well known classical design problem is discussed. However limitations to the classical theory of this problem potentially limit its application to certain real world problems by virtue of the small number of precision points and unspecified order. A prototype of the problem was designed and simulated for visualizing a four bar link (4BL – simple movable closed chain linkage) mechanism engineering problem using MATLAB software. The aim is to facilitate the analysis, dynamic simulation of the four-bar mechanisms. First, a brief review of some of the current computer-aided learning software for four-bar mechanisms is presented. These software packages provide two-dimensional visualization and computational capabilities necessary to synthesize and analyze four-bar mechanisms. The Chapter also reviews the kinematics of four-bar mechanisms as they pertain to their geometric modelling followed by the design approach of the graphical MATLAB simulation for 4BL mechanisms. A preliminary validation test using pre and postest questionnaires and focus group discussion with student participation in using the MATLAB simulation has facilitated in visualizing the engineering concepts of 4BL mechanism at UNITEN.

Chapter 5 presents the subject of spur gear design problem in mechanical engineering (ME) that deserves significant attention since the subject relies heavily on the combination of complex mathematics formulas, graphs and tables. Many students are facing difficulty in understanding the subject of spur gear design in Universiti Tenaga Nasional (UNITEN). Hence, a study was conducted to enhance the learning experience of UNITEN's ME students in the subject matter. In this research the deterministic, divide and conquer algorithms were employed to improve the problem solving technique. The combination of these algorithms provided room for students to make assumption based on their input rather than having the computer to do all the calculation and as well as to break down the complicated problems into sub problems. Additionally it was designed to examine the effectiveness of using computer-aided software to improve the learning experience and comparing it to the conventional approach. The data collection procedures involved using both methods of teaching, which is the Computer-Aided Learning (CAL) approach and conventional approach. Both groups of students were assessed based on their prior computer experience, preferred learning styles (as a result of Felder-Solomon's ILS questionnaire), learning experience and comparing the result in solving spur gear design problem with and without the CAL software. An analysis of the participants' results revealed that CAL software has helped ME students in solving spur gear design problem with a better approach. The first evaluation finds that more students from the group with Spur Gear Analysis Software were able to create a more cost and weight effective spur gear design problem as compared with the group without Spur Gear Analysis Software. The second evaluation finds that visual representation such as graph helps ME students in visualizing the concept and foreseeing the relationship between each parameter involved, making their learning experience easier and better.

Chapter 6 of the book discusses the design and model of an AR based software that uses visual cues to enhance the learning process and the outcome perception results of the cues. Augmented Reality (AR) is a potential area of research for education, covering issues such as tracking and calibration, and realistic rendering of virtual objects. The ability to augment real world with virtual information has opened the possibility of using AR technology in areas such as education and training as well. In the domain of Computer

Aided Learning (CAL), researchers have long been looking into enhancing the effectiveness of the teaching and learning process by providing cues that could assist learners to better comprehend the materials presented. Although a number of works were done looking into the effectiveness of learning-aided cues, but none has really addressed this issue for AR-based learning solutions.

In Chapter 7, the preferences of UNITEN students in using augmented reality technology user interface applications to solve selected engineering problems are compared. Cross sectional study design and Wilcoxon-Signed Rank Test approach was adopted to analyze the difference in the rankings of the engineering applications. Within a controlled environment, this research further applies "Trials within the same session but with breaks between tasks" an affirmed reliable method in measuring learnability that has been rarely explored by any related works locally The results were captured through descriptive statistical analysis. The findings provided reliable evidences that multiple function user interface (MFIT) is more effective than the tangible user interface (TUI) for engineering students. Simultaneously, this research also presents evidences that MFIT is better than TUI in the engineering problem solving statistically.

Chapter 8 presents a novel AR application using touch to print technique. Many learning methods has changed the way students learn. One method that is achieving much attention is augmented reality (AR). AR is a technology that blends simulated and real environment during the learning, interaction and visualization process. As such, an AR ATAPS with a new interaction technique (touch-to-print) was designed, tested and evaluated. The aim was to provide an improved user interface i.e. without having to use markers so as the learner could focus more on the visualization process. The AR ATAPS is capable of recognizing the 4BL mechanisms (based on Grashof's law and user input data) i.e. drag-link, crank-rocker, double-rocker and parallelogram linkage and have been used in this study as an adjunct to traditional problem solving method. The touch to print interaction technique which is the main contribution of this research, has been, useful to engage the user in the problem with a new interactive and learning experience as compared to the previous method i.e. the use of markers to interact with a virtual object. The interaction technique method uses seven functions that are recognizable by the ATAPS (rotational, link colour change A, B, C, D, pause and voice command) for the user to touch the symbols on the paper and the system to model and analyze accordingly in real-time 3D environment. This study explores how far AR technology has come to support students in their learning and interest in using this technology. The objective of this Chapter was to determine the usefulness of touch to print interaction user interface for an AR application. A hands on practical lab was conducted with first year engineering students at UNITEN. The evaluation and effectiveness of the AR ATAPS as an alternative to textbooks and current software learning packages was examined by means of a single-institutional evaluation study using mainly statistical quantitative techniques and ANOVA analysis. The prospective study (total sample size = 30) at University Tenaga Nasional (UNITEN) validated aspects of AR ATAPS interaction technique and provided feedback on the interface design and its problem solving method. The results of the study showed that most of the participants never been experienced with AR applications before, but the ideas of implementing AR as a simulation tool for learning the kinesthetic and dynamic subjects is well accepted with a very beneficial feedback. Based on the findings it was found that there is a positive changing in terms of the visualizing and imagining of the four-bar linkage mechanisms (4BL) which lead to a good understanding of this subject. Further development of AR applications in the learning environment is being discussed.

Teaching conceptual and qualitative material effectively while leveraging the contents efficiently has been an elusive goal for many computer aided learning (CAL) packages in the past. With the advent of newer technologies such as multimedia and virtual reality these technologies are being researched and applied to various areas of educational settings, especially in science and technology. However the potential of these technologies has not been fully exploited, particularly in the teaching of engineering.

In this book, an innovative approach based on the principle of CAL is used to design and implement integrated packages known as Advanced Technology Assisted Problem Solving or ATAPS packages, which guide students step-by-step to complete various Engineering Mechanics problems.

In summary, this book is concerned with the design, development, and evaluation of a new form of interactive visual multimedia CAL problem solving packages which replace traditional problem-solving aspects of undergraduate level of teaching selected mechanical engineering subjects. These packages, subsequently named as ATAPS packages, provide the instructors with an economical means of facilitating the teaching of engineering concepts in Dynamics to a large population of undergraduates. This study seeks to examine the overall effectiveness of these packages that provide an integration of traditional teaching (lectures) with technology assistance (software packages) to enhance the students learning in today's resource limited environment.

It is difficult to address every aspect of the technologies employed in the development of ATAPS packages because of the rapid change, upgrades and evolvement of hardware and software. Nevertheless, hopefully, this book will provide useful and rich description through a reflective analysis of the technologies employed in this study. Alternative approaches are reviewed. This could provide a deeper insight into methods for educational research.

Acknowledgment

Trademark Notice

3D Studio Max™ is a trademark of Autodesk, Inc.
Maya™ is a trademark of Alias, Inc.
Cinema4D™ is a trademark of Maxon, Inc.
Lightwave™ is a trademark of NewTek, Inc.
Director™ is a trademark of Adobe, Inc.
Authorware™ is a trademark of Macromedia, Inc.
Flash™ is a trademark of Adobe, Inc.
ToolBook™ is a trademark of Asymetrix, Inc.
Photoshop™ is a trademark of Adobe, Inc.
Illustrator™ is a trademark of Adobe, Inc.
Mathlab™ is a trademark of MathWorks, Inc.

I would like to thank IGI Global for granting me the opportunity to publish this book and share my knowledge with academicians, educationist, and individuals who wish to broaden their horizons about educational technologies pertaining to engineering education. Availability of the vast and untapped potentials of the latest technologies inspired me to integrate their benefits in educational software tools. In addition, I would like to express my special thanks to:

Research students Dr. Waleed Maqableh, Dr. Lee Chen Kang, Dr. Kather Ali, Ms. Jee Geak Ying, and Ms. Nur Maisarah Shahril Khuzairi.

Research assistants Mr. Abdul Hadi Arham and Mr. Mohd Anwar Muslimen.

E-Science Funding – from the Ministry of Science, Technology and Innovation (MOSTI) Malaysia.

Universiti Tenaga Nasional – for all kinds of support to complete the research.

Prof. Dr. Richard M. Felder and Barbara A. Solomon for the use of the Index of Learning Styles Questionnaire.

My wife Dr. Kirandeep Kaur Sidhu for proof reading the book. This publication would not have been possible without her.

Many thanks to the team at Idea Group Publishing, in particularly Joel Gamon, who have answered my queries, provided advice, and support.

I also like to thank all the referees who provided constructive and comprehensive reviews of the contents of this book. They work anonymously in the background, but their tireless effort is much appreciated.

3D Studio Max™ is a trademark of Autodesk, Inc.

Maya™ is a trademark of Alias, Inc.

Cinema4D™ is a trademark of Maxon, Inc.

Lightwave™ is a trademark of NewTek, Inc.

Director™ is a trademark of Adobe, Inc.

Authorware ™ is a trademark of Macromedia, Inc.

Flash™ is a trademark of Adobe, Inc.

ToolBook ™ is a trademark of Asymetrix, Inc.

Photoshop ™ is a trademark of Adobe, Inc.

Illustrator ™ is a trademark of Adobe, Inc.

Mathlab ™ is a trademark of MathWorks, Inc.

Chapter 1
Problems Faced by Educators and Students in Teaching and Learning Engineering

ABSTRACT

In this chapter, the author presents the problems faced by educators and students in teaching and learning engineering courses. In general, teaching of engineering courses are relatively challenging due to the nature of the courses that are perceived as "difficult courses" by a number of students. This course is built upon the strong fundamental knowledge in physics and mathematics. This course requires the students to have a strong abstract thinking, reasoning, and problem-solving skills. Due to this problem, a further research study (a continuation from previous study) was embarked at the National Energy University, which is also known as Universiti Tenaga Nasional (UNITEN), Malaysia. Selected lecturers and students were randomly selected and interviewed to find the reasons for this problem.

INTRODUCTION

The purpose of this research was to explore the problems faced by mechanical engineering (ME) lecturers teaching the course and students studying difficult topics i.e. four link bar (4BL) mechanisms, shaft crank, motions such as linear, projectile and curvilinear, spur gear design and rotational of x and y axes. In response to the new paradigm shift in engineering education that emphasizes

DOI: 10.4018/978-1-7998-0465-9.ch001

on the use of Information and Communication Technologies (ICT) to facilitate the teaching and learning for tertiary education, there have been numerous studies that proposed the solutions utilizing the Computer Aided Learning (CAL) platform. The usability of the CAL software to facilitate the learning of the students need further investigation empirically. In this research, several advanced technology assisted problem solving (ATAPS) packages which can be considered as a branch of CAL was enhanced with simulations and augmented reality (AR) technologies. Additionally, this research was also aimed at identifying the students' attitude towards outcome based education (OBE). Literature has shown that OBE could enhance students' understanding and make the learning more meaningful. The theoretical framework of this study was based on selected learning styles instruments and software measurement inventory such as Honey and Mumford, Felder and Silverman, the Ogden's Personality and Learning Styles Questionnaire. The research study further investigated the usability of the enhanced ATAPS packages. Through the usability testing, two of the usability evaluation instruments namely (i) system usability scale (SUS) and (ii) post-study system usability questionnaire (PSSUQ) were employed which will be further discussed in other Chapters of this book.

One of the major key to serve the nation's development is the technological advancement especially Information and Communication Technologies (ICT). Apart from this, systematic knowledge of engineering (gained through appropriate structures) signifies a major role in the attainment of a high level of technological advancement. Most developing countries however faces difficulty to impart adequate knowledge and training to engineers at different levels of education due to lack of supporting infrastructure and modern technological teaching aids.

According to Lee (2018), in the era of knowledge driven society, changes occur rapidly throughout the industries and the marketplace. Two main factors that drive the rapid change in the market environment and the society are the globalization and the revolution of ICT. Knowledge has played an important role as a sustainable competitive factor for survival in the dynamic marketplace (Bhatti et al., 2016; Hana, 2013; Li & Liu, 2014). As mentioned by the well-known management expert, Peter Drucker (1993), the main challenge in the knowledge-based economy is how to make the information and knowledge productive enough to compete in this constantly changing environment. The globe is moving towards the era of post-industrial knowledge society where the future will be essentially determined by the ability to utilize knowledge for unique ideas, products and services that emphasize on innovation efforts for competitive advantage. In fact, the shift from material and labor intensive products and processes to knowledge intensive products and services are the unavoidable major trends in the knowledge driven economy (Ferriera et al., 2017; Sala et al., 2016).

The paradigm shift in engineering education in response to the revolution of the knowledge society has created new opportunities for education practitioners. The utilization of ICT to facilitate the teaching and learning in engineering has increasingly raised the interest of the education practitioners throughout the world. As can be identified from the literature, students evolving from the classroom that are integrated with the multimedia environment are more motivated and enthusiastic in learning activities and therefore more committed to their relationship to knowledge (Bai et al., 2016; Doong & Ho, 2012; Koh et al., 2015; Wang et al., 2014). There are many computer-based learning packages in the domain of engineering (e.g. the animation and simulation software, virtual laboratories, experimental video, virtual reality application, simulation based virtual laboratories and etc.) which have been developed for higher learning institutions and also used commercially in western countries (Cheng, 2014; Fang & Guo, 2016; Kojmane et al., 2016; Morsi et al., 2015; Pocsova et al., 2016; Re & Giubergia, 2013; Salam, 2011).

The rapid growth of technology has also made our life easier in one way or another such as the implementation of technology in education which is also known as Computer Aided Learning (CAL), Computer-Based Learning (CBL), and Computer Assisted Learning. The introduction of CAL has improved the conventional teaching and learning environment as CAL allows information or knowledge to be presented via multimedia such as image, text, audio and video (Sidhu et al., 2011). A study by Felder and Silverman (1988), has mentioned that the information presented in such way could cater for a diverse learning style, so students could benefit the most out of it. Hence, CAL is not something new and has also been implemented into many education fields which require more than just words to explain and present the theory or concept such as engineering education. In general, CAL has been used widely in the field of engineering as the subject involves heavily on the combination of mathematical formulas, free body diagrams and the effect of forces, where CAL is quoted by researchers as a solution to create a better problem-solving environment for engineering students. In the context of Malaysia, the development of TAPS (Technology Assisted Problem Solving) packages which is considered as a new branch of CAL founded by Professor Manjit has made one of the pioneering efforts to facilitate the engineering problem solving in the domain of mechanical engineering (Lee & Manjit,

2015). The TAPS packages are aimed at coaching students, particularly who need additional support in applying principles presented in lectures to problems, on the best approach to solve a particular engineering problem in a step-by-step or logical approach (Manjit, 2006). The knowledge and information discussed and published in this book would further build on top of the previously developed TAPS packages to propose new Advanced Technology Assisted Problem Solving (ATAPS) packages and with tested effective patterns of interactions that aim to provide better interactions to aid in the engineering problem solving tasks and visualization.

Hence this book highlights recent works done in ATAPS packages and engineering education but specifically on solving mechanical engineering problems. This work was also carried out to find research possibilities that could refine ATAPS to not only solve problem but also to improve the engineering student's learning experience (i.e. making assumptions / analysis / predictions and visualization).

BACKGROUND

Since the introduction of the Outcome Based Education (OBE) into the engineering programs by the Board of Engineers Malaysia (BEM) in 2009, all engineering programs in Malaysia such as civil engineering, electrical and electronic engineering and as well as Mechanical Engineering program have to be OBE compliance in order to obtain full accreditation from BEM. Hence, Bloom's Taxonomy was developed in 1956 by Benjamin Bloom to measure and categorize the Outcome-Based Education (Eisner, 2000).

With the emergence of problems in teaching and learning, studies by Sidhu (2010), Neumann, Neumann and Hood (2011) and Smetana and Bell (2012) suggested that CAL can solve the engineering problems and foresee it as an alternative solution. This is because CAL has the ability to address different learning styles with its information presented in multimedia, 2-D and 3-D mechanic models environment, creating an interactive learning environment in teaching engineering subjects which require real-time visuals to demonstrate the concepts. Works done by Huang et al. (2011), Raveendran and Batmanabane (2012), Sidhu and Ali (2013), Vattam et al. (2014) and Sanchez at al. (2014) who incorporated CAL into the engineering learning environment have provided evidence demonstrating the fact that CAL outperforms the conventional learning method in term of student's learning outcomes. However, the existing CAL solution still does not promote

Figure 1. Relationship between OBE, bloom taxonomy and learning curve

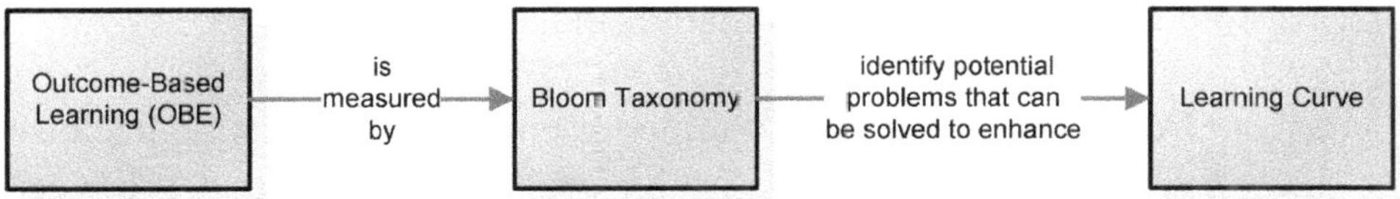

student's learning outcome to a higher level thinking which is level 4, 5 and 6 (based on the Bloom's Taxonomy model). The findings gathered from the above authors have raised a research opportunity to create a comprehensive problem solving environment, so that engineers are equipped with a higher thinking skill upon graduation.

There have not been much studies conducted in the Malaysian context in using TAPS packages to enhance the understanding of engineering students in some difficult and challenging mechanical engineering topics such as mechanics dynamics, spur gear design, rotational of x and y axes, four link bar mechanisms (4BL), and crank shaft just to name a few. Therefore, it has led to this research, study of ATAPS to create a comprehensive learning environment to promote higher thinking skills in the mentioned perspective.

The introduction of Bloom's Taxonomy in evaluating the OBE (Figure 1) has made problem in teaching and learning to be more noticeable. This is because Bloom's Taxonomy allows the engineering undergraduate's learning outcomes to be categorized or quantified into the 6 levels (Level 1 – Remembering, Level 2 – Understanding, Level 3 – Applying, Level 4 – Analyzing, Level 5 – Evaluating, Level 6 – Creating). The potential pitfall and problems identified can be solved to enhance the learning curve of engineering students.

In this research, the ATAPS packages are designed in relation with the constructivism learning theory where learners will have a personal discovery of knowledge, discovering the relationships between concepts in the engineering topics mentioned earlier in this book. The learning is viewed as a result of mental construction and each level of it can be categorized by Bloom Taxonomy. Although the ATAPS has improved the learning experience, but now the question that educators are concern is the ability of students to solve problems with higher thinking skills which is Level 4, 5 and 6 in Bloom's Taxonomy.

Reviewing on past studies, it has raised several problems in teaching and learning with higher thinking skills in the engineering subject, especially machinery design in mechanical engineering. This is because it involves heavily on the combination of mathematical formulas, graphs, tables and the effect of forces. Each problem statement is elaborated as follows:

- Most CAL packages are not suitable for engineering learning because they are merely an interface just for problem solving that produces simple answers without showing the full workings. It does not support the engineering learning in term of visualizing, making assumption and better understanding the concept which is essential to achieve higher level thinking skills. As such, further work is required to develop a more comprehensive approach that could serve this purpose and task.
- Problem among engineering students in achieving Bloom's Taxonomy higher level thinking skills (Level 4, 5 & 6) has been identified to be one of the major issues found among engineering students (Mohayidin et al., 2008). Previous CAL work showed has been too lenient in tackling this matter where attention is given more to problem solving, whereas student's ability in attaining a higher thinking skill is neglected. The researcher believes that higher thinking skill is equally important in problem solving too in designing CAL packages to produce engineering graduates that are not only academically but also technically competent.
- ATAPS in mechanical engineering's subjects has yet to be applied to wider Malaysian audiences and globally. Thus, there is inquisitiveness of how Malaysian engineering undergraduates would respond to ATAPS in mechanical engineering subjects, which will be helpful in improving the ATAPS approach in Malaysian context.

STUDENTS PROBLEMS IN LEARNING ENGINEERING

This Chapter focuses on the problems faced by students in the field of mechanical engineering, in particularly the engineering mechanics dynamics course. This subject is chosen because a number of academicians in the field as reported in the literatures found that the main problem faced by students is visualization of dynamic motion of particles or rigid bodies.

In general the first year mechanical engineering programme offers a number of subjects that are mix of mathematics, formulas, schematic diagrams and problem solving subjects. Based on examination results of local Malaysian universities, this course is considered relatively "difficult subject" for the students to grasp. This supports the evidence from the previous years that most moderate achievers failed to provide complete answers for the mechanics dynamics and structural questions. There were candidates who tried to guess the answers, especially names of the motions, structures and components. The low achievers, on the other hand, failed to answer multiple choice questions and to list down structures and components. For the structural questions, majority of them failed to do the calculation and they drew the diagrams wrongly. There were few students who did not attempt to answer at all.

Additionally, mistakes in simplifying the formula have caused some high-achievers to provide the wrong answers. For the moderate achievers, only a few of them were able to answer completely based on the instruction. For example, most candidates only wrote the formulae without showing the calculation. Mistakes also occurred when they wrote the wrong units/formulas of the calculation. For the low achievers, their main problem was they did not know how to solve the problem and the majority would just guess the answer.

Katarzyna (2002), reported that the problems many undergraduate students face while studying the engineering mechanics dynamics course is the difference in understanding with regard to what is being taught in the classroom. Undergraduate students often expect a variety of teaching methods to be used in their learning. Although, the lecture method is a common way of delivering knowledge to students, it treats all students on the same level of the basic acquired knowledge. However, most of the students do not bring the same academic preparation (do not have the same foundation, motivation, interest, ability to learn/grasp) and come from different disciplines with varying learning styles. This results in different starting points, progress rates, and ultimately different levels of satisfaction, academic progress, and performance.

Some entry-level undergraduate students find certain engineering subjects difficult to understand and this discourages learning from taking place. As a result of this problem, if the lectures are too fast, this group of students may not be able to keep pace with the rest of the class thus the gap in their knowledge will only get wider as compared to the more brilliant students. In this situation, the instructors are forced to find alternative methods (for example conducting extra classes) to help these students in understanding the subject matter. Since some students may take more time to understand the problem solving techniques and may require the lesson to be repeated several times before they understand, it is proposed to employ and use multimedia and VR technologies to help them understand the engineering problems.

A major challenge facing instructors in teaching the Engineering Mechanics Dynamics course at UNITEN, Malaysia is in helping students relate the theory to the physical world. Past experiences in teaching the first and second year students indicated that there were students who found it difficult to visualize some of the concepts in engineering and to apprehend theory and practical. In general, engineering textbooks cannot represent mechanical actions such as movement of linkages, pistons, and crankshaft in the form of dynamic illustrations (animated forms). Since these mechanical actions are very important in visualization experiments, the motions and actions of these objects maybe worthwhile to be shown in animated forms. Therefore, it can be clearly seen that educating students in engineering related field is made difficult by the complex ideas and phenomena demonstrated by conventional methods.

However one of the enormous challenges that any institution looks forward to employ new technologies is the training of instructors to integrate these new technologies in their teaching of engineering. In general, instructors must be computer literate to develop and utilize multimedia and mixed reality technologies such as virtual reality (VR) and augmented reality (AR). Therefore, properly trained instructors are required if multimedia, VR and AR technologies are to be integrated and implemented within the classroom. In this book the term slow learners has been used for those students who have difficulties in understanding Engineering Mechanics Dynamics subject only and does not reflect on their general academic performance.

EVOLUTION OF CAL

The first use of computers by educational institutions and the introduction of computers in classroom teaching and technical training began in the 1950's (Megarry, 1983). According to Robert (1994), the most pervasive tool to deliver education is the computer. This is probably due to the increasing popularity of personal computers (Law and Maguire, 1993). In the 1960's and 1970's, more teaching devices (computers) and teaching CAL applications were developed for teaching and job training. As a result, improvements in computer design during the 1960's enabled the integration of text and graphics; a move that encouraged the development of computer based training programs. Such use was, however, limited, as programming was costly, slow and tedious, and the computing relied on a mainframe, or central computer, to do all the work. The release of the first microcomputers during the late 1970's assisted computing in becoming more portable and affordable.

The paradigm shift from textbook learning to CAL started in the 1980's. Ease of use took a quantum leap with the introduction of the Apple Macintosh computer in 1984. Operation of the computer and its programs became more intuitive and user-friendly largely as a result of intense research in the area of human-computer interaction where graphical user interface was introduced (Kinshuk, 1996). The low cost of hardware equipment and educational software motivated the use of CAL in education and training. Today, as affordable personal computers become more powerful, the range of operations that can be reasonably performed has increased rapidly.

Role of New Technologies in CAL

The use of newer technologies in CAL packages i.e. multimedia and virtual reality, allows highly engaging activities such as interactivity and simulation. Interactive multimedia systems for learning came into existence in the early 90's (Robert, 1994). According to Negroponte (1995), multimedia is simply a mixture of data on digital basis. Multimedia systems support the physical and logical coexistence and interactive use of mixed media classes such as print, audio, and video in specific application environments. Cairncross and Mannion (1999) stated that interactive multimedia systems have the potential to create high quality learning environments that actively engage the learner. Additionally, Cairncross (2002) pointed out that the key elements of multiple media, such as user control over delivery of information and interactivity, could be used to enhance the learning process.

Another emerging technology that is being used in CAL is virtual reality (VR). VR systems were first introduced in the learning environment in mid 90's (Macpherson, 1998). The term 'virtual reality' is currently used to describe a range of computer-based systems in which a user can explore hardware or software generated 'micro world' (artificial environments) that allow close resemblance to reality. VR extends the interaction-oriented features of multimedia i.e. modeling objects and their behaviors in virtual environments, integrating position-tracked human-computer interaction devices, and performing numerically intensive computations for real-time navigation.

The prime feature of VR is 'visualization' followed by 'interactivity'. Special VR hardware and software are thus required to allow human-computer interaction to permit input of the user's actions and movement to the computer, and to provide corresponding simulated feedback to the user.

An early application of such system was the flight simulator used to train pilots. However, it is in the area of hi-tech computer games that many of the application developments in this field have taken place.

Although VR has been used for educational purposes (Bell and Scott, 1995, Dede *et al.*, 1996, and Kim *et al.*, 2001), the potential of VR is just beginning to be exploited by a few science and engineering educators (Manseur, 2005 and Liarokapis *et al.* 2007). This will be discussed further in Chapter 2.

Problems Associated With CAL

Despite many successes, CAL packages are bound by numerous limitations and drawbacks. Schank (1994) reported that early educational software offered tedious drill (exercise programs) and were primarily built on a set of pre-defined steps that did not engage the students in their learning process and thus failed to promote learning. According to Kinshuk (1996), the pre-defined steps incorporated into the computer software presented knowledge to learners through a specific order in a text form. The knowledge of a CAL package did not go beyond the information stored in its memory and has likely done little to improve learning. When using educational CAL packages, students typically rely on the package for feedback. Therefore, the prime challenges of a CAL package is to response to students' needs and provide appropriate feedback so that the students can understand the theories and be able to apply them in solving problems. However, according to Kinshuk (1996), the rigid and unresponsive nature made the instructional effectiveness less than satisfactory.

After the introduction of Artificial Intelligence (AI) to CAL, AI researchers in the 1970's started to transfer expertise from human professionals to a machine accessible form and then redirected this knowledge back to other human beings (Barr and Feigenbaum, 1982). AI researchers attempted to organize the problem-solving skill and expertise in a way that was suitable for teaching and consulting (Kinshuk, 1996). To develop such an intelligent teaching package, AI techniques such as knowledge representation, user modeling, and natural language processing have widely been incorporated into intelligent tutoring systems (ITS) (Kinshuk, 1996).

From the perspective of human computer interaction, the research on CAL is very focused on how to represent the learning content and tends to neglect the impact of the user-interface in the learning process (Guttormsen *et al.*, 2000). On the other hand, Cairncross and Mannion (2001) pointed out that there is a growing evidence that interactive multimedia is not being fulfilled. The author further argued that early designs were driven by technology (i.e. focusing mainly on physical interface) rather than pedagogy.

Additionally, Cairncross (2002) reported that it could take 40 hours to develop one hour's worth of quality interactive multimedia learning. This lengthy development time is extremely expensive because it is difficult and costly to find human experts to develop the interactive multimedia CAL packages. As an option with most other teaching media, instructors wanting to use interactive multimedia CAL package in their teaching can choose to develop simple unsophisticated materials locally or purchase more sophisticated, and thus more expensive, teaching materials from professional development units (commercial software house, etc).

Schank (1994) argued that most multimedia programs are not suitable for learning because they merely add video and graphics to page-turning programs. Kinshuk and Patel (2003) added that the collection of multimedia objects i.e. pictures, graphics, sounds, and video does not guarantee proper learning, especially when the complexity of the task, skill or learning increases. According to Bell (2002), in the learning process paradigm, it's not just to ensure the transfer of learning but what is learned should make a difference. As such, further work is required to develop more effective approaches that could serve this purpose and task.

At the simplistic level, some multimedia programs and software have limited qualities and have been dubbed *'electronic page turners'* (Goldman and Torrisi-Steele, 2004). Interaction is limited to the student clicking a mouse to sequentially take them to the next page. To fully develop the enormous potential of multimedia as an interactive learning package, multimedia elements and activities must possess a multitude of interactivity option made available to the student.

Subsequently, a range of problems limit the use of hi-end hardware and software in the educational environments. These problems are mainly associated to VR technology. The main problem of VR technology in the educational environments is associated with high cost. Although prices are rapidly decreasing, immersive VR systems still cost hundreds of thousands of dollars (Rizzo *et al.*, 1997, Klett, 2002). Alternatively, cheaper options for VR technology are also available such as desktop virtual reality which only requires special browsers to interact and visualize the virtual objects Other problems that are restricting the pursuit of the VR vision in the educational environments include the limited availability of applications development expertise and software licenses for multiple users (learners).

Application of CAL in Engineering Education

In general, education, in higher learning institutions in countries with developing economics still focuses on older educational models of linear progression or surface learning, whereas counterparts from developed nations provide predominantly high-impact audio-visual perception.

The western countries, particularly the UK and USA have used computers and CAL packages to motivate students of higher learning institutions since the 1960s (Ismail, 2001). Although most higher learning instutitions would employ and use new technology in teaching, several academicians commented that they still lack widespread practical experience of developing multimedia-learning materials (Julia *et al.*, 2002).

However since the emergence of newer hardware and software technologies for multimedia and VR, educational practitioners began to research further on the pedagogical effectiveness of these technologies. In a developing country such as Malaysia, multimedia technology was first briefly introduced in the late 1990s and became popular with the launch of Multimedia Super Corridor (MSC) (Norhayati *et al.*, 2001). To achieve the substance of Vision 2020, the government has set up a blueprint for the MSC. The MSC is a massive 750-square-kilometer high-tech information zone encompassing the Kuala Lumpur City Center (KLCC), Putrajaya (administrative center) and Kuala Lumpur International Airport (KLIA). To spearhead the development of the MSC and give shape to its environment, seven initiatives for multimedia applications have been identified. These initiatives are borderless marketing, smart schools, electronic government, multi-purpose card, telemedicine, research and development world wide manufacturing webs. Of these, the smart school initiative is regarded by the previous Prime Minister as a specific response to Malaysia's need to make the critical transition from an industrial economy to a knowledge-based society (Mahathir Mohammed, 1998). Subsequently VR hardware and software are being used in various research fields such as medicine, manufacturing and for scientific visualization.

The nation is devoting the abovementioned massive MSC to create the perfect environment for companies and education sector wanting to develop, distribute, and employ multimedia products and services. One of MSC's primary areas of multimedia applications includes "*smart schools*" where educational software packages are being customized to facilitate teaching and student learning purposes in primary and secondary schools. In general, although the educational sector is aware of the presence of MSC, these new technologies (multimedia and VR) are hardly being researched further for its utility and potential in teaching engineering. Additionally, multimedia and VR systems are not available on a large scale to support learning environments. Since these technologies are in its infancy state especially in the higher learning institutions further research is needed to address its usefulness and benefits.

Already locally commercialized CAL packages (custom made packages developed in Malaysia) are available for learning purposes, but these CAL packages vary in subjects and are more suitable for primary and secondary school levels. There are still many areas that need improvement because these CAL packages are simply computerized textbooks with page turning architecture that do not promote learning (lack of capability to motivate user to participate actively in the learning process). Additionally, most commercial CAL packages are based on outdated theories of learning. For example, long series of short answer questions are provided to students in the CAL packages but the CAL packages are not intelligent enough to provide appropriate reasons and solutions to the incorrect answers given by the students (most commonly alert messages i.e. '*correct*' or '*wrong*' are used).

Thus to improve the learning process, particularly in higher learning institutions, suitable CAL packages tailored to the needs of students to enhance learning should be employed and be made an integral part of the current curriculum of the educational system. However, in order for CAL to become an effective learning platform, certain criteria must be met. For example, the instructors should be prepared to accept and encourage students to use the technology as an additional learning aid in an effective manner. The full potential of CAL cannot be realized if computers are merely used by students for preparing laboratory reports and assignments.

In general the development of CAL packages is a tedious process. Even simple graphics and animations require significant effort to develop the contents of CAL packages that employs multimedia and VR technologies.

For engineering and technology education, CAL multimedia and VR applications can include computer simulation, numerical analysis, computer aided design (CAD), computer aided manufacture (CAM), and electronic communications (Palmer, 2000).

Although many CAL learning environments have emerged in general, in this book the term technology assisted problem solving (TAPS) packages will be used to refer to the advanced packages that were developed in this study and used by students to assist them in their learning. The main difference between conventional CAL and TAPS packages is that TAPS packages do not include text of theories learnt by students during normal classroom lectures but rather constituted selected engineering problems that are difficult for students to understand from text books. The TAPS packages aimed at coaching students, particularly who need additional support in applying principles presented in lectures to problems, on the best approach to solve a particular engineering problem in step-by-step or logical approach. Although a TAPS package may be considered as a form of CAL, more specifically, TAPS packages were designed to allow independent problem solving, develop logical thinking, and promote learning of the subject matter. At present, the developed TAPS packages are not a complete set of engineering problems that could be used throughout an entire semester but it is envisaged that more TAPS packages could be developed if these are proven to be effective. TAPS packages are discussed in details by Manjit (2009).

ADVANCED TECHNOLOGY ASSISTED PROBLEM SOLVING PACKAGES (ATAPS)

This book is concerned with the design, development and evaluation of low cost computer based problem solving aids to deliver knowledge in engineering courses. The discussion and case studies of the book is based solely on the development of the TAPS packages that focuses on the study in Mechanical Engineering at University Tenaga Nasional, Malaysia.

The aim of the work reported in this book was to enhance the understanding and problem solving skills of students (slow learners) taking the Engineering Mechanics Dynamics course at University Tenaga Nasional (UNITEN), Malaysia. This was done by analyzing the learning styles, student needs, development of computer based problem solving packages, and exposure to modern technologies. The main development path chosen was to introduce and employ new educational computer technologies.

Initially, the perceived need was to use computer based learning aids to show selected engineering problems involving motion of bodies and to provide a convenient platform for students to visualize and study them.

More specifically, the objectives of the study explained in this book can be summarized as follows:

1. To evaluate whether slow learners could better visualize and solve engineering problems with the aid of advanced technology assisted problem-solving (ATAPS) packages.
2. To explore how best to assist students to understand engineering concepts via interactive and virtual environment.
3. To determine the best approach in integrating the various elements in the ATAPS packages with the aim of promoting learning and understanding of engineering concepts suitable for slow learners.

Whilst the aim was to design and develop ATAPS packages for the projection of selected engineering problem solving tasks of the subject matter in a virtual environment, the study:

1. Limits its scope to the topics in the selected Engineering Mechanics Dynamics problems.
2. Employs a multi approach learning and problem solving environment in engineering i.e. 2-D, 3-D, coach based and desktop virtual reality for student learning (particularly slow learners).
3. Customizes the ATAPS packages to facilitate learning of Dynamics course at UNITEN, Malaysia.

This study began with an initial hypothesis that the design of interactive multimedia problem-solving applications for engineering could be improved through the development of productive computer packages and employment of some principles of good user interface design.

Initially multiple user interface designs were planned with the aim to determine how multimedia attributes could be exploited to enhance problem solving. One of the important issues that the initial user interface design raised was to integrate multimedia attributes into the engineering problem solving steps leading from problem statement to solution which is fundamental. The study further explored this issue.

The book describes a new form of CAL that is termed as advanced technology assisted problem-solving (ATAPS) packages. The key concepts of ATAPS package, user interface, and contributing technologies that have been used to develop the packages are discussed in details in Chapter 2. The ATAPS packages are different from other CAL engineering packages in the sense that they provide multiple design approaches to solve selected Engineering Mechanics Dynamics problems in interactive 2-D and 3-D environment, static and dynamic illustrations, coach, desktop virtual reality (DVR) and simple intelligence, so that a student can visualize the engineering principles. Each package was developed as a separate component to solve different engineering problems. This new design approach to learning, visualizing, and problem solving in engineering provides significant concepts to the development of the ATAPS packages.

OUTLINE OF THE BOOK

An introduction of this study and an outline of the problems and objectives are given in Chapter 1. Computer aided learning (CAL) is discussed in brief regarding its evolution and the role of new technologies. Problems associated with CAL, application of CAL in engineering education and problems encountered by engineering students are discussed in depth.

Theories and models of instructional design are discussed in brief in Chapter 2. Various cognitive styles have been described in the literature and many different measurement tools are available for these styles. Chapter 2 of this book presents a summary of these learning styles and measurement methods. The Chapter also discusses the appropriateness of Felder-Solomon's Learning Style Questionnaire (Felder and Silverman, 1988) and Ogden's personality and learning styles questionnaire in the context of computer-aided learning and the categorization of people according to their styles.

Chapter 3 of this book reviews different approaches in the user interface design and examines the role of interface design in various educational environments. The designing of the user interface has been discussed in details. This Chapter also reviews some problems in traditional user interface designs and lists some guidelines for better user interface design. Additionally, the role of the interface in educational environments and student's preferences towards interfaces in learning environments is briefly discussed.

In Chapter 4, the synthesis and analysis of four link bar (4BL) mechanisms problem which is well known classical design problem is discussed. However limitations to the classical theory of this problem potentially limit its application to certain real world problems by virtue of the small number of precision points and unspecified order. A prototype of the problem was designed and simulated for visualizing a four bar link (4BL – simple movable closed chain linkage) mechanism engineering problem using MATLAB software. The aim is to facilitate the analysis, dynamic simulation of the four-bar mechanisms. First, a brief review of some of the current computer-aided learning software for four-bar mechanisms is presented. These software packages provide two-dimensional visualization and computational capabilities necessary to synthesize and analyze four-bar mechanisms. The Chapter also reviews the kinematics of four-bar mechanisms as they pertain to their geometric modelling followed by the design approach of the graphical MATLAB simulation for 4BL mechanisms. A preliminary validation test using pre and postest questionnaires and focus group discussion with student participation in using the MATLAB simulation has facilitated in visualizing the engineering concepts of 4BL mechanism at UNITEN.

Chapter 5 shows that the subject of spur gear design problem in mechanical engineering (ME) deserves significant attention since the subject relies heavily on the combination of complex mathematics formulas, graphs and tables. Many students are facing difficulty in understanding the subject of spur gear design in Universiti Tenaga Nasional (UNITEN). Hence, a study was conducted to enhance the learning experience of UNITEN's ME students in the subject matter. In this research the deterministic, divide and conquer algorithms were employed to improve the problem solving technique. The combination of these algorithms provided room for students to make assumption based on their input rather than having the computer to do all the calculation and as well as to break down the complicated problems into sub problems. Additionally it was designed to examine the effectiveness of using computer-aided software to improve the learning experience and comparing it to the conventional approach. The data collection procedures involved using both methods of teaching, which is the Computer-Aided Learning (CAL) approach and conventional approach. Both groups of students were assessed based on their prior computer experience, preferred learning styles (as a result of Felder-Solomon's ILS questionnaire), learning experience and comparing the result in solving spur gear design problem with and without the CAL software. An analysis of the participants' results revealed that CAL software has helped ME students in solving spur gear design problem with a better approach. The first evaluation finds that more students from the group with Spur Gear Analysis Software were able to create a more cost and weight effective spur gear design problem as compared with the group without Spur Gear Analysis Software. The second evaluation finds that visual representation such as graph helps ME students in visualizing the concept and foreseeing the relationship between each parameter involved, making their learning experience easier and better.

Chapter 6 of this book describes an Augmented Reality (AR) application. AR is a potential area of research for education, covering issues such as tracking and calibration, and realistic rendering of virtual objects. The ability to augment real world with virtual information has opened the possibility of using AR technology in areas such as education and training as well. In the domain of Computer Aided Learning (CAL), researchers have long been looking into enhancing the effectiveness of the teaching and learning process by providing cues that could assist learners to better comprehend the materials presented. Although a number of works were done looking into the effectiveness of learning-aided cues, but none has really addressed this issue for AR-based learning solutions. This Chapter discusses the design and model of an AR based software that uses visual cues to enhance the learning process and the outcome perception results of the cues.

In Chapter 7 of this book, the preferences of UNITEN students in using augmented reality (AR) technology user interface applications to solve selected engineering problems are compared. Cross sectional study design and Wilcoxon-Signed Rank Test approach was adopted to analyze the difference in the rankings of the engineering applications. Within a controlled environment, this research further applies “Trials within the same session but with breaks between tasks” an affirmed reliable method in measuring learnability that has been rarely explored by any related works locally. The results were captured through descriptive statistical analysis. The findings provided reliable evidences that multiple function user interface (MFIT) is more effective than the tangible user interface (TUI) for engineering students. Simultaneously, this research also presents evidences that MFIT is better than TUI in the engineering problem solving statistically.

Chapter 8 discusses an advanced novel AR technology assisted problem application for engineering any learning methods has changed the way students learn. Since AR is a technology that blends simulated and real environment during the learning, interaction and visualization process, a new AR ATAPS with novel interaction technique (touch-to-print) was designed, tested and evaluated. The aim was to provide an improved user interface i.e. without having to use markers so as the learner could focus more on the visualization process. The AR ATAPS is capable of recognizing the 4BL mechanisms (based on Grashof’s law and user input data) i.e. drag-link, crank-rocker, double-rocker and parallelogram linkage and have been used in this study as an adjunct to traditional problem solving method. The touch to print interaction technique which is the main contribution of this research, has been, useful to engage the user in the problem with a new interactive and learning experience as compared to the previous method i.e. the use of markers to interact with a virtual object. The interaction technique method uses seven functions that are recognizable by the ATAPS (rotational, link colour change A, B, C, D, pause and voice command) for the user to touch the symbols on the paper and the system to model and analyze accordingly in real-time 3D environment. This study explores how far AR technology has come to support students in their learning and interest in using this technology. The objective of this Chapter was to determine the usefulness of touch to print interaction user interface for an AR application. A hands on practical lab was conducted with first year engineering students at UNITEN. The evaluation and effectiveness of the AR ATAPS as an alternative to textbooks and current software learning packages was examined by means of a single-institutional evaluation study using mainly statistical quantitative techniques and ANOVA analysis. The prospective study (total sample size = 30) at University Tenaga Nasional (UNITEN) validated aspects of AR ATAPS interaction technique and provided feedback on the interface design and its problem solving method. The results of the study showed that most of the participants never been experienced with AR applications before, but the ideas of implementing AR as a simulation tool for learning the kinesthetic and dynamic subjects is well accepted with a very beneficial feedback. Based on the findings it was found that there is a positive changing in terms of the visualizing and imagining of the four-bar linkage mechanisms (4BL) which lead to a good understanding of this subject. Further development of AR applications in the learning environment is being discussed.

In summary, this book is concerned with the design, development, and evaluation of a new form of interactive multimedia CAL problem solving packages which replace traditional problem solving aspects of undergraduate level of teaching selected mechanical engineering subjects. These packages, subsequently named as ATAPS packages, provide the instructors with an economical means of facilitating the teaching of engineering concepts in Dynamics to a large population of undergraduates. This study seeks to examine the overall effectiveness of these packages that provide an integration of traditional teaching (lectures) with technology assistance (software packages) to enhance the students learning in today's resource limited environment.

REFERENCES

Bai, Y., Mo, D., Zhang, L., Boswell, M., & Rozelle, S. (2016). The impact of integrating ICT with teaching: Evidence from a randomized controlled trial in rural schools in China. *Computers & Education*, *96*, 1–14. doi:10.1016/j.compedu.2016.02.005

Barr, A., & Feigenbaum, E. A. (1982). *The Handbook of Artificial Intelligence, 2*. Los Altos: Kaufmann.

Bell, C. R. (2002). Managers as Mentors: Building Partnerships for Learning. Berrett-Koehler Publishers.

Bell, J. T., & Scott, H. F. (1995). The Investigation and Application of Virtual Reality as an Educational Tool. *Proceedings of the American Society for Engineering Education Annual Conference*, 1718 – 1728.

Bhatti, W. A., Larimo, J., & Coudounaris, D. N. (2016). The effect of experiential learning on subsidiary knowledge and performance. *Journal of Business Research*, *69*(5), 1567–1571. doi:10.1016/j.jbusres.2015.10.018

Cairncross, S. (2002). *Interactive Multimedia and Learning: Realizing the Benefits* (PhD Thesis). Napier University.

Cairncross, S., & Mannion, M. (1999). How multimedia functions in engineering education. *Engineering Science and Education Journal*, *8*(3), 100–107. doi:10.1049/esej:19990302

Cairncross, S., & Mannion, M. (2001). Interactive Multimedia and Learning: Realizing the Benefits. *Innovations in Education and Teaching International*, 156 – 164.

Cheng, G. (2014). Exploring students' learning styles in relation to their acceptance and attitudes towards using Second Life in education: A case study in Hong Kong. *Computers & Education*, *70*, 105–115. doi:10.1016/j.compedu.2013.08.011

Dede, C., Salzman, M., & Loftin, B. (1996). The Development of a Virtual World for Learning Newtonian Mechanics. In P. Brusilovsky, P. Kommers, & N. Streitz (Eds.), *Multimedia, Hypermedia, and Virtual Reality*. Berlin: Springer. doi:10.1007/3-540-61282-3_10

Doong, S. H., & Ho, S.-C. (2012). The impact of ICT development on the global digital divide. *Electronic Commerce Research and Applications*, *11*(5), 518–533. doi:10.1016/j.elerap.2012.02.002

Eisner, E. W. (2000). *Benjamin Bloom 1913-99*. Retrieved from http://www.ibe.unesco.org/publications/ThinkersPdf/bloome.pdf

Fang, N., & Guo, Y. (2016). Interactive computer simulation and animation for improving student learning of particle kinetics. *Journal of Computer Assisted Learning*, *32*(5), 443–455. doi:10.1111/jcal.12145

Felder, R. M., & Silverman, L. K. (1988). Learning and Teaching Styles in Engineering Education. *Journal of Engineering Education*, *78*(7), 674–681.

Ferreira, F., Faria, J., Azevedo, A., & Marques, A. L. (2017). Product lifecycle management in knowledge intensive collaborative environments: An application to automotive industry. *International Journal of Information Management*, *37*(1), 1474–1487. doi:10.1016/j.ijinfomgt.2016.05.006

Goldman, J. D., & Torrisi-Steele, G. (2004). Constructive Pedagogies of Interactivity on a CD-ROM to Enhance Academic Learning at a Tertiary Institution. *International Journal of Educational Technology*.

Guttormsen, S., & Krueger, H. (2000). Using New Learning Technologies with Multimedia. *IEEE MultiMedia*, 40–51.

Hana, U. (2013). Competitive Advantage Achievement through Innovation and Knowledge. *Journal of Competitiveness*, *5*(1), 82–96. doi:10.7441/joc.2013.01.06

Huang, E. Y., Lin, S. W., & Lin, S. (2011). A quasi-experiment approach to study the effect of e-mail management training. *Computers in Human Behavior*, *27*(1), 522–531. doi:10.1016/j.chb.2010.09.021

Ismail, J. (2001). The design of an E-learning system: Beyond the hype. *The Internet and Higher Education Journal*, *4*(3-4), 329–345. doi:10.1016/S1096-7516(01)00069-0

Julia, M., Niall, S., & Su, W. (2002). *Report on the British Council Seminar and Asian Association of Open Universities Exhibition in Malaysia*. Retrieved from the World Wide Web from http://cvu.strath.ac.uk/niall/malaysia/

Katarzyna. (2002). *Knowledge Representation, Content Indexing and Effective Teaching of Fluid Mechanics Using Web-Based Content* (Masters Thesis). Massachusetts Institute of Technology.

Kim, J., Heon, P. S., Tae, Lee, H., & Yuk, K. C. (2001). Virtual Reality Simulations in Physics Education. *Interactive Multimedia Electronic Journal of Computer-Enhanced Learning*, *3*(2).

Kinshuk, R. (1996). *Computer Aided Learning for Entry accountancy students* (PhD Thesis). De Montfort University.

Kinshuk, R., & Patel, A. (2003). *Optimizing Domain Knowledge Representation with Multimedia Objects. In Learning and Teaching with Technology - Principles and practices*. London: Kogan.

Klett, K. (2002). Improving Engineering Student Learning in a Web-Based Learning Space Due to Virtual Reality Techniques and Advanced Interactivity. *Proceedings of the 2002 eTEE Conference Switzerland*, 195 – 101.

Koh, J. H. L., Chai, C. S., Benjamin, W., & Hong, H.-Y. (2015). Technological Pedagogical Content Knowledge (TPACK) and Design Thinking: A Framework to Support ICT Lesson Design for 21st Century Learning. *The Asia-Pacific Education Researcher*, *24*(3), 535–543. doi:10.100740299-015-0237-2

Kojmane, J., & Aboutajeddine, A. (2016). Enjoyeering Junior: A hands-on activity to enhance technological learning in an engineering dynamics course. In *2016 International Conference on Information Technology for Organizations Development (IT4OD)* (pp. 1–6). IEEE. 10.1109/IT4OD.2016.7479267

Law, R. C. H., & Maguire, R. B. (1993). A critical survey of intelligent tutoring systems. *Congressus Numerantium*, *92*, 55–64.

Lee, C. K. (2018). *Patterns of Interactions for Technology Assisted Problem Solving (TAPS) Packages* (PhD thesis). Universiti Tenaga Nasional (UNITEN).

Lee, C. K., & Manjit Sidhu, S. (2015). Engineering Students Learning Preferences in UNITEN: Comparative Study and Patterns of Learning Styles. *Journal of Educational Technology & Society*.

Li, D., & Liu, J. (2014). Dynamic capabilities, environmental dynamism, and competitive advantage: Evidence from China. *Journal of Business Research*, *67*(1), 2793–2799. doi:10.1016/j.jbusres.2012.08.007

Liarokapis, F., Mourkoussis, N., White, M., Darcy, J., Sifniotis, M., Petridis, P., ... Lister, P. F. (2007). Augmented Reality and Web 3D to Support Engineering Education. *Proceedings of the Engineering and Technology Education Conference*, 24-28.

Macpherson, C. (1998). Virtual Reality: What is the state of play in education? *Australian Journal of Educational Technology*, *14*(1), 60–74.

Mahathir, M. (1998). *Multimedia Super Corridor. Subang Jaya. Pelanduk Publications (M)*. Sdn. Bhd.

Manjit Sidhu, S. (2006). *Development and applications of interactive multimedia TAPS (Technology Assisted Problem Solving) packages for engineering* (PhD Thesis). University Malaya, Malaysia.

Manjit Sidhu, S. (2009). *Technology-Assisted Problem Solving for Engineering Education: Interactive Multimedia Applications*. Idea Group Engineering Science Reference & Publishing.

Manseur, R. (2005). Virtual Reality in Science and Engineering Education. *35th ASEE/IEEE Frontiers in Education Conference*, 8 – 13.

Mbonu, O. J. (2000). Practical training of engineering personnel—challenges ahead. *Proceedings of COREN 10th Engineering Assembly*, 30–38.

Megarry, J. (1983). *Thinking, Learning and Educating: The role of the computer. In Computers and education* (pp. 15–28). Nichols Pub. Co.

Morsi, R., & Mull, S. (2015). Digital Lockdown: A 3D adventure game for engineering education. In 2015 IEEE Frontiers in Education Conference (FIE) (pp. 1–4). IEEE. doi:10.1109/FIE.2015.7344072

Negroponte, N. (1995). *Being digital*. New York: Alfred A. Knopf.

Neumann, D. L., Neumann, M. M., & Hood, M. (2011). Evaluating Computer-Based Simulations, Multimedia and Animations that Help Integrate Blended Learning with Lectures in First Year Statistics. *Australasian Journal of Educational Technology*, *27*(2), 274–289. doi:10.14742/ajet.970

Norhayati, A. M., Halimah, B. Z., & Siew, P. H. (2001). *Courseware Development to Motivate Life-Long Reading Habits. In Multimedia at Work* (pp. 76–81). IEEE.

Palmer, S. (2000). On-and off-campus computer usage in engineering education. *Computers and Education Journal, 33*(3), 141 – 154.

Peter, F. (1993). *Drucker*. Post-Capitalist Society.

Pocsova, J., Mojzisova, A., & Takac, M. (2016). Application of the visualization techniques in engineering education. In *2016 17th International Carpathian Control Conference (ICCC)* (pp. 596–601). IEEE. 10.1109/CarpathianCC.2016.7501166

Raveendran, R., & Batmanabane, G. (2011). Alternatives to Animals in Teaching: Experience in an Indian Medical School. *Proceedings of the ALTEX 8th World Congress*, 355-360.

Ré, M. A., & Giubergia, M. F. (2013). Virtual Laboratory for a first experience in dynamics. In *ICBL2013 – International Conference on Interactive Computer aided Blended Learning* (pp. 239–244). Academic Press.

Rizzo, A. A., Buckwalter, J. G., & Neumann, U. (1997). Virtual reality and cognitive rehabilitation: A brief review of the future. *The Journal of Head Trauma Rehabilitation*, *12*(6), 1–15. doi:10.1097/00001199-199712000-00002

Robert, D. (1994). *The Development, Formative and Summative Evaluation of a Computer Multimedia Tutorial: A Case Study* (PhD thesis). Walden University.

Sala, A., Landoni, P., & Verganti, R. (2016). Small and Medium Enterprises collaborations with knowledge intensive services: An explorative analysis of the impact of innovation vouchers. *R & D Management*, *46*(S1), 291–302. doi:10.1111/radm.12196

Salam, S. (2011). *Investigation of students' learning dynamics in an augmented reality environment to support individualisation, collaboration and creativity*. Academic Press.

Schank, R. C. (1994). Active Learning through Multimedia. *IEEE MultiMedia*, *1*(1), 69–78. doi:10.1109/93.295270

Sidhu, M. S. (2010). *Technology-Assisted Problem Solving for Engineering Education: Interactive Multimedia Applications*. Hershey, PA: Engineering Science Reference. doi:10.4018/978-1-60566-764-5.ch004

Sidhu, M. S., & Ali, K. (2013). Computer Aided Simulation of a 4BL Engineering Problem Using Matlab. *Australian Journal of Basic and Applied Sciences*, *7*(10), 381–387.

Sidhu, M. S., & Kang, L. C. (2011). Interactive augmented reality environments for engineering with open systems. In *2011 IEEE Conference on Open Systems* (pp. 1–5). IEEE. 10.1109/ICOS 2011.6079279

Smetana, L. K., & Bell, R. L. (2012). Computer Simulations to Support Science Instruction and Learning: A critical review of the literature. *International Journal of Science Education*, *34*(9), 1337–1370. doi:10.1080/09500693.2011.605182

Vattam, S. S., Goel, A. K., Rugaber, S., Hmelo-Silver, C. E., Jordan, R., Gray, S., & Sinha, S. (2011). Understanding Complex Natural Systems by Articulating Structure-Behavior-Function Models. *Journal of Educational Technology & Society*, *14*(1), 66–81.

Wang, S.-K., Hsu, H.-Y., Reeves, T. C., & Coster, D. C. (2014). Professional development to enhance teachers' practices in using information and communication technologies (ICTs) as cognitive tools: Lessons learned from a design-based research study. *Computers & Education*, *79*, 101–115. doi:10.1016/j.compedu.2014.07.006

KEY TERMS AND DEFINITIONS

ATAPS Packages: A new branch of CAL software with advanced functions to facilitate the engineering problem solving in the domain of mechanical engineering. The ATAPS packages are aimed at coaching students, particularly who need additional support in applying principles presented in lectures to problems, on the best approach to solve a particular engineering problem in a step-by-step or logical approach.

Augmented Reality (AR): Is an interactive experience of a real-world environment where the objects that reside in the real-world are enhanced by computer-generated perceptual information, sometimes across multiple sensory modalities, including visual, auditory, haptic, somatosensory, and olfactory.

OBE: Outcome-based education is an approach that focuses on outcomes such as the achievements of students that are measurable, proven and can be improved. It is simply the establishment of expected goals or outcomes for different levels of elementary-secondary education, and a commitment to ensure that every student achieves at least those minimum proficiencies before being allowed to graduate.

TAPS Packages: A branch of CAL software to facilitate the engineering problem solving in the domain of mechanical engineering. The TAPS packages are aimed at coaching students, particularly who need additional support in applying principles presented in lectures to problems, on the best approach to solve a particular engineering problem in a step-by-step or logical approach.

Virtual Reality (VR): is the use of computer technology to create a simulated environment. Unlike traditional user interfaces, VR places the user inside an experience. Instead of viewing a screen in front of them, users are immersed and able to interact with 3D worlds.

Chapter 2
ICT and Education:
A Review

ABSTRACT

This chapter reviewed the literature on the trends and paradigm shifts in engineering education. Through the literature, the problems in mechanical engineering specifically for the learning of mechanics dynamics were identified. The importance of information and communication technologies (ICT) and the application of computer-aided learning (CAL) in engineering education was discussed. In the last section of this chapter, the theory of learning styles associated with engineering education was studied.

ENGINEERING EDUCATION

Engineering education is regarded as one of the important educational domains in the tertiary education. As defined by Cheshier (1998);

"Engineering education for the professional focuses primarily on the conceptual and theoretical aspects of science and engineering aimed at preparing graduates for the practice of engineering closest to the research, development, and conceptual design functions" (p. 36).

In the earlier days, when the engineering education was introduced, it followed the apprenticeship form with an emphasis on hands-on practical education (ASEE, 1987). Later, engineering education was formalized into

DOI: 10.4018/978-1-7998-0465-9.ch002

the academic studies and followed a general pattern of teaching and learning. In general, engineering education involves two distinct learning environments which are the classroom teaching on theoretical knowledge for conceptual understanding and the laboratory sessions to obtain the practical knowledge (Balamuralithara & Woods, 2009).

Recently, the paradigm shift in engineering education raised the attention of the engineering communities and is actively discussed in numerous reports (Auguestine, 2005; Froyd et al., 2012; National Science Board, 2007; Prados, 1998; Wince-Smith, 2005). As discussed by Duderstadt (2008), one of the characteristics for new paradigms of engineering education is the change of pedagogical style that shifted from classroom based pedagogy to active learning approaches that engaged problem-solving skills and team building, by which it is more focused on discovery oriented, interactive and collaborative learning experiences. Table 2.1 summarizes the characteristics for old and new paradigms of engineering education as discussed in details by Duderstadt (2008).

Froyd et al. (2012) identified the five major shifts which have reshaped (the first and second shifts) or currently reshaping (the third, fourth and fifth shifts) the engineering education for the past 100 years. The details are summarized in Table 2.

In response to the aforementioned paradigm shift in engineering education, the engineering communities from academic, research and industry have contributed invaluable efforts. One of the efforts that the engineering community focused on is the utilization of information and communication technology (ICT) to foster the teaching and learning in engineering education (Manjit, 2006; Froyd et al., 2012). Although educational technology is not an ultimate solution for the paradigm change in engineering education, but at least, educational technology supports some of the attributes in the new engineering education paradigm such as support for educational methods that stress active learning; communication, team work, group problem-solving skills; cater to different learning styles and foster more effective learning for engineering education. The issues related to ICT and its impact on education are discussed in section "Computer-Aided Learning (CAL) for Engineering Education."

Table 1. Comparisons on the characteristics for old and new paradigms of engineering education

Engineering Education (characteristics)	Old Paradigms	New Paradigms
The Curriculum	▪ Focused on scientific and technical courses as the core of an engineering education.	▪ Not only focused on scientific and technical courses but include new curriculum that must reflect a broad range of concerns on economic, political, social, and environmental context of engineering practice.
The Ability and Skills	▪ Technical knowledge and skills	▪ Technical knowledge and skills ▪ Communication skills ▪ Teamwork / Teambuilding ▪ Ability to adapt to an increasingly diverse world ▪ Ability not only to adapt to change but to actually drive change ▪ Social and environmental consciousness
Depth vs. Breadth	▪ Follow the reductionism approach - focused on teaching and scholarship on increasingly narrow and specialized topics.	▪ Focused more on comprehensive curriculum and broader educational experience in which topics are better connected and integrated. ▪ Holistic approach to address social needs
Pedagogical Style	▪ Classroom based pedagogy – lecture-dominated system (large lecture courses, rigidly defined problem assignments, highly structured laboratory courses).	▪ Active learning approaches that engage problem-solving skills and team building. ▪ Focused on discovery-oriented, interactive, and collaborative learning experiences.
Lifelong Learning	▪ Less awareness on lifelong learning – concerns more on the acquiring of knowledge for future jobs.	▪ Aware on the importance of lifelong learning and concerns more on the knowledge of how to learn and continue to learn throughout the life time.
New Technologies	▪ From microscopic level of info-bio-nano.	▪ To the macroscopic level of global systems.
A Broader Concern	▪ Focus primarily on educating students for the engineering profession.	▪ Educating not simply professional engineers but a new breed of graduates with an engineering-based, liberal education.

Mechanical Engineering and Mechanics Dynamics

Mechanical engineering is defined as the discipline that applies the principles of engineering, physics, and materials science for the design, analysis, manufacturing, and maintenance of mechanical systems (The American Heritage Dictionaries, 2011). Mechanical engineering is a diverse subject that derives its breath from the need to design and manufacture everything from small individual parts and devices (e.g., micro scale sensors and inkjet

Table 2. Five major shifts in engineering education (Adopted from Froyd et al., 2012)

	Engineering Education Major Shifts	Descriptions
1	A shift from hands-on and practical emphasis to engineering science and analytical emphasis	- Engineering curricula moved from hands-on, practice-based curricula to ones that emphasized mathematical modeling and theory-based approaches.
2	A shift to outcomes-based education and accreditation	- Based largely on the actions of the Accreditation Board for Engineering and Technology (ABET), engineering education and accreditation became outcomes based. - Through ABET and its predecessors, accreditation has provided quality control for engineering education, seeking to assure that graduates of accredited programs are prepared for professional practice.
3	A shift to emphasizing engineering design	- This major shift emphasis on professional practice that enabling students to move from being passive viewers of engineering action to taking their place as active participants or creators within the field of engineering.
4	A shift to applying education, learning, and social-behavioral sciences research	- Use of educational objectives, and, more broadly, student learning outcomes and Bloom's Taxonomy (original or revised) is an indication that research in psychology, education, and learning science is having a noticeable influence on the engineering education community. - Student engagement or involvement in learning is the second area for which there is evidence of the influence of research in psychology, education, and learning science on the practice of engineering education. - Inquiry and inquiry-based or guided approaches focus first on the question, problem, challenge, or goal to be addressed. Then, students learn content, concepts, and processes while addressing the question, problem, challenge, or goal. - Understanding by Design (UbD) is an increasingly popular tool for educational planning focused on teaching for understanding. The emphasis of UbD is on backward design, the practice of first looking at the outcomes in order to design curriculum units, performance assessments, and classroom instruction. - Evidence of increased emphasis on a broader range of knowledge, skills, and attributes (or habits of mind and modes of thinking) for engineering graduates abounds. - Emergence of increased emphasis on a scholarly approach to engineering education is indicated.
5	A shift to integrating information, computational, and communications technology in education	- Emphasized application of information, communication, and computational technologies (ICCT) in engineering education. - Some of the principal instructional technologies and their applications have been: content delivery (television, videotape, and the Internet); programmed instruction (individualized student); feedback; personal response systems; computational technologies; intelligent tutors (second phase of individualized); student feedback; simulations; games and competitions; remote laboratories; grading.

printer nozzles) to large systems (e.g., spacecraft and machine tools) (Hone, 2012). It is one of the oldest and popular engineering disciplines. According to The Accreditation Board for Engineering and Technology (ABET), "*the program must demonstrate that graduates have the ability to: apply principles of engineering, basic science, and mathematics (including multivariate calculus and differential equations) to model, analyze, design, and realize*

Table 3. Competencies that students are expected to acquire in traditional teaching methods, students learn within a classroom with the help of lectures, notes or by using textbooks

By completing the Dynamics course, students should be able to:
To understand the purpose of dynamics in relation to engineering.
To understand the basic physical concepts of dynamics.
To create a free body model of a rigid dynamical system.
To perform a kinematic and kinetic analysis of a rigid body particle or system.
To analyse fundamental rigid body dynamics problem using concepts of work, energy, impulse and momentum.
To apply Newton's Laws to particles and rigid bodies to solve problems related to dynamic behaviour.
To use properly the theory and equations and practice them in practical applications.
To know the appropriate use of modern computational tools related to mechanical dynamics.
To understand basic dynamics as used in the design process.

physical systems, components or processes; and work professionally in both thermal and mechanical systems areas."(ABET, 2013).

Generally, mechanics dynamics is offered as one of the important subject to the first year or second year mechanical engineering students. In the respective course, students discover mechanisms and are often asked to solve problems related to modeling, analysis and design of system in motion (Kojmane & Aboutajeddine, 2016). In the literature, syllabi of the course of dynamics of different universities around the world describe the competencies that students are expected to acquire are summarized in Table 3.

Educational research reveals that such type of a passive environment is not effective and results in limited knowledge retention by students (Tembe & Kamble, 2016). In the traditional (instructor-centered) engineering teaching, where the instructor introduces each topic by lecturing the general principle followed by derivation of mathematical models and illustrative applications of the models, may lead to the students difficulties in understanding the concepts and principles (Haque, Nayna, & Ahmed, 2016). In addition, many students think that dynamics is a collection of problem-specific tricks instead of a unified body of knowledge built upon a very limited number of basic equations and principles (Barroso, Morgan, & Simpson, 2008). It can be observed that many of the students would treat each course as an individual entity that has minimal connections to each other. Without seeing the relevance of the material, they tend to forget the materials instantly after the final exam. This poses a challenge for knowledge transfer to other courses later in their degree program. As such, following basic pedagogies issues have been identified as underlying the difficulties most students have (Barroso et al., 2008): 1) forgetting, misconceptions and misapplication of prior knowledge leading to difficulties with knowledge transfer between courses; 2) difficulties developing models and connecting the response of those models to real system behavior; 3) critical thinking about complex problems and systems, both in how to break down a problem and identify appropriate simplifying assumptions, as well as how to evaluate their problem solution and system behavior. According to Kojmane & Aboutajeddine (2016), the dynamics course is critical for students where they can be exposed to engineering real-world problems, and can sense the real meaning of engineering and design. The example given by Kojmane & Aboutajeddine (2016), "*for instance, in the automotive and aeronautical industry, employers become more and more requiring as regards the quality of the hired engineers, and there's an augmenting demand for engineers who master the design tools of systems in motion, and who are capable to solve problems related to the development of products.*"

Therefore, to deal with the work environment of today's engineers, it is necessary to adapt the engineering education curriculum, in order to produce engineering graduates well prepared for their role as competent engineers and well equipped with the critical in-demand job skills (Nasr, 2014). Engineering teachers are often encouraged to incorporate new instructional strategies such as active learning methods that motivate their students to be more actively engaged with the subject course materials (Tembe & Kamble, 2016). As described by Bonwell (1991), in active learning, students participate in their own learning through: a) taking responsibility in the learning process and b) involving in activities beyond just listening – such as reading, writing, discussing and solving problems. Active learning is found/believed to improve the students' overall learning in engineering education. Through the research findings by Prince (2004), active learning consists of students being engaged in learning through participating in knowledge construction process (cognitive process), resulting in a deeper knowledge. According to Prince (2004), strong evidence is available to support the use of active learning for improvement in recall of information. There are many techniques / activities that can involve the engineering students in active learning. Active learning techniques include in-class group work, concept mapping, use of multimedia in course materials, project based learning, problem-based learning, think-pair-share strategy and "clicker" questions using student response systems (Aglan & Ali, 1996). Technology can be used to promote active learning and meaningful learning and it provides interactivity between students and

Table 4. ICT and its functions in support for the formal teaching and learning (Adopted from Kirkwood, 2009)

Function	Descriptions
Presentation	Making materials and resources (text, data, sounds, still and moving images, etc.) available for students to refer to, either at predetermined times or 'on demand'.
Interaction	Enabling learners to actively engage with resources, to manipulate or interrogate information or data, and so on.
Dialogue	Facilitating communication between teachers and learners or between peers for discussion, cooperation, collaboration, and so on.
Generative activity	Enabling learners to record, create, assemble, store and retrieve items (text, data, images, etc.) in response to learning activities or assignments and to evidence their experiences and capabilities.

knowledge content (Shapiro, 1998) (Tembe & Kamble, 2016) (Wang, Shen, Novak, & Pan, 2009).

ICT AND ITS ROLE IN EDUCATION

The interest of using new technologies among the researchers to improve the teaching and learning in education is increasing rapidly (Martin et al., 2011). The Information and Communication Technology (ICT) which includes the major digital technologies (software, hardware and networking technologies) can be considered as one of the great potential tools to facilitate the educational change and enhance the learning process (Dhandabani & Sukumaran, 2014; Jonassen, 1999; Smeets, 2005; Tinio, 2003). In fact, the impact of ICT in education can be identified through various educational levels from pre-school, primary education, secondary education to tertiary education. In tertiary education, as argued by Kirkwood (2009), ICT was usually adopted to support the formal teaching and learning process through the following ways as shown in Table 4.

Using ICT to foster better learning environments build on the foundation of educational models to facilitate the teaching and learning process. The popular educational models used in tertiary education are summarized as shown in Table 2.5. ICT greatly facilitates the implementation of all types of educational learning theories and delivery models that support and encourage innovative teaching and learning modalities both for theoretical courses and lab courses (Dhandabani & Sukumaran, 2014). Meanwhile, there is a great potential for utilization of ICT in education as students in tertiary education

Table 5. Educational models used in teaching and learning (Adopted from Dhandabani & Sukumaran, 2014)

Educational Practices	Purpose
Curriculum development Theories and Principles (Connelly & Xu, 2012)	Utilized in creating course curriculum and syllabi.
Instructional System Design (Kelting-Gibson, 2013)	Utilized in creating the overall course learning plan that comprises of course rationale, objectives, schedule, assessments, evaluation and feedback.
ADDIE Model (Welty, 2007)	Utilized in designing teaching materials, student leaning worksheets, assessment materials, learning evaluation materials, feedback materials.
Bloom Taxonomy (Bloom, 1969)	Utilized in designing a micro plan of a session and objectives of the session.
Cone of Learning Model (Dale, 1969)	Utilized in designing self-learning activities for learning simulation.
Kirkpatrick's Model (Kirkpatrick & Craig, 1976)	Utilized for training evaluation.
Assessment Model (Kay & Knaack, 2008)	Utilized for learning assessment.
Feedback Model (Pendleton, 1984)	Utilized for collection of online feedback.
Robert Gagne's Model (Gagne, 1985)	Utilized for blended learning in classroom/laboratory teaching.
Epistemological Rationale (Tsai, Chai, Wong, Hong, & Tan, 2013)	Utilized to examine educational assessments
Training and Educational Psychology (Ross, Morrison, & Lowther, 2010), (Lindsay, 2007)	Utilized in providing guidance and counseling.

today are known as 'digital natives'; those who grew up with technology all around them (cell phones, laptops, and tables) (Morsi & Mull, 2015).

ICT has been used to foster better learning environments since 1970s (Deniz & Cakir, 2006; Sakamotao et al., 1979). According to Reiser (2001), there were three stages of the evolution for educational technology and can be identified between 1900 till 2000s. The stages refer to the visual instruction movement (1900 -1920s), the audiovisual movement (1920s - 1970) and the computer instruction movement (1970s onwards). Technology evolved throughout the evolution (see Figure 1). Reiser provided detailed descriptions of the evolution of educational technology and further derived a conclusion from the history that the media technologies had minimal impact on instructional practices. However, he predicted positively that over the next decade, the ICT will contribute to greater changes in instructional practices than any previous media technology.

Figure 1. The evolution of educational technology (1900s – 2000s) (Lee & Manjit Sidhu, 2013)

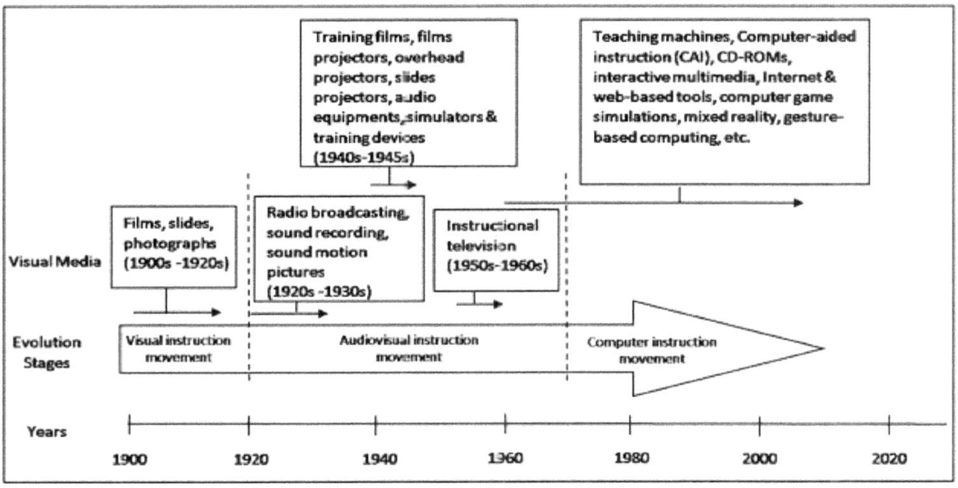

When the significant computer revolution started, the influence of the computer was best seen in its multimedia configuration which included an integration of multiple media elements such as text, graphics, animation, audio and video into a coherent learning environment; which in turn transformed student learning and problem solving approach (Entwistle et al., 1992; Janson, 1992). The benefits of utilizing the multimedia technologies to support and assist in learning can be easily identified in the literature. For example, Kulik and Kulik (1991) argued that computer assisted learning (CAL) including multimedia is an effective form of instruction because it produces high student outcomes of achievement in short periods of time. Furthermore, research studies on multimedia learning found that multimedia foster cognitive change (Mautone & Mayer, 2001); deepens perception of ideas and of concepts (Achikalin & Dur 2005); improved performance in problem solving (Zheng et al., 2006); and facilitate information processing in learning (Mayer & Moreno, 2003; Rieber & Kini, 1991). In addition, the engagement for the learners in their learning process can be improved through the aid of multimedia in learning. This is because the use of multimedia elements in an integrated way involved a wider range of stimuli, verbal and visual, thereby increasing the state of engagement (Teoh & Neo, 2006; Zheng et al., 2008).

In today's knowledge society and innovation-led economy, higher order thinking skills are important as a response to the quick and uncertain change in the market environment. Examples of higher order thinking skills are critical system thinking, decision making, and problem solving (Miri et al., 2007). In fact, the higher order thinking skills correspond to the higher scale of Bloom's taxonomy which includes the analysis, synthesis and/or evaluation stages (Kek & Huijser, 2011; Miri et al., 2007; Zohar, 2004). It is important to develop the higher order thinking skills in students especially during the tertiary education stage. As known, the ultimate goal of higher education is to promote a deep approach to learning (higher levels of learning) (Biggs, 2003). As a continuous effort in improving education, many education practitioners actively propose a variety of learning approaches that may contribute to develop higher order thinking skills. In this situation, educational technologies used for cognitive support in learning may contribute to enhance the students' higher-order thinking skills (Kirschner & ErKens, 2006; McMahon, 2009; Ringstaff & Kelley, 2002).

Another significant achievement for the use of ICT to enhance teaching and learning is the formation of student-centered learning environments. As described by Hannafin and Land (1997), student-centered learning environments emphasize on "*constructing personal meaning by relating new knowledge to existing conceptions and understandings; technology promotes access to resources and tools that facilitate construction*" (p.170). From a constructivist perspective learners need to be encouraged to take responsibility for their learning, with learners having a sense of ownership of learning experiences (Squires, 1999). In a student-centered learning environment, the students will have higher learner control throughout the learning process. The term learner control has been defined and used to represent choice in learning in various contexts, such as learning content, method of delivery, order and way of learning, time and pace of learning as well as the level of learning instructions or activities (Stemler, 1997). Such conditions can be implemented with the aid of multimedia technology to create a suitable learning environment that focuses on student-centered. This is fundamental because multimedia supports the characteristics by which the delivery of information (through multiple media / format), the organization in which it is delivered and the timing of that delivery can be controlled by the user (Cairnscross & Mannion, 1999; Manjit, 2006). In addition, the non-linear fashion offered by the multimedia provides flexibility in navigation (Cairnscross & Mannion, 1999; Manjit, 2006). As such, the learners may have the opportunity to choose on the learning contents that suit to their learning levels accordingly. Within the learning contents, the hypertext and hypermedia functionalities provide the flexibility for the learners to further explore the information that are connected and related to each other for further explanations and reference.

Another important attribute of multimedia is the interactivity capability. Borsook and Higginbotham-Wheat (1991), believed that the computer's interactive potential makes it unique in the history of educational/instructional technology and sets it apart from all other instructional devices. As stated by Manjit et al. (2004), interaction refers to the reciprocal action of two phenomena and has both a physical connotation (one entity operating on another) and a psychological connotation (two entities influencing each other's behaviour). By having interactivity, the learning environment generated by computer will respond instantly and accordingly based on the learners' response. The key attributes model of multimedia (Manjit et al., 2004) is shown in Figure 2. As such, multimedia provides the opportunity to fulfill the aforementioned conditions that it can support for flexible control and aid in facilitating the student-centered learning environments. By allowing flexibility in control, greater individualization in learning at different paces can be achieved (Teoh & Neo, 2006). This will lead to a greater opportunity for the student to learn at his or her full potential (Tway, 1995) and nurture "deep learning" (Kabouridis, 2010). Deep learning is associated with the higher scale of Bloom's taxonomy which includes the analysis, synthesis and evaluation.

COMPUTER-AIDED LEARNING (CAL) FOR ENGINEERING EDUCATION

Computer-aided learning (CAL) is defined as the means of attaining knowledge via computer-based technologies, leading to a lasting change in behavior (Huczynski & Johnston, 2005). CAL environments increasingly are using a combination of interactive multimedia and virtual reality (VR) features in delivering the learning contents (Manjit, 2006). The research studies on CAL started a few decades ago. Although the research results vary, the tendency is towards the positive side of support for students learning (Deniz & Cakir, 2006). Within engineering education, there are many computer-based learning packages which have been developed for higher learning institutions and also used commercially in western countries (Manjit, 2006). However Manjit (2006), argued that these computer-based learning packages may not be suitable for the use of local students as the dialect used may hinder / delay

Figure 2. Key attributes model of multimedia (Adopted from Manjit, 2006)

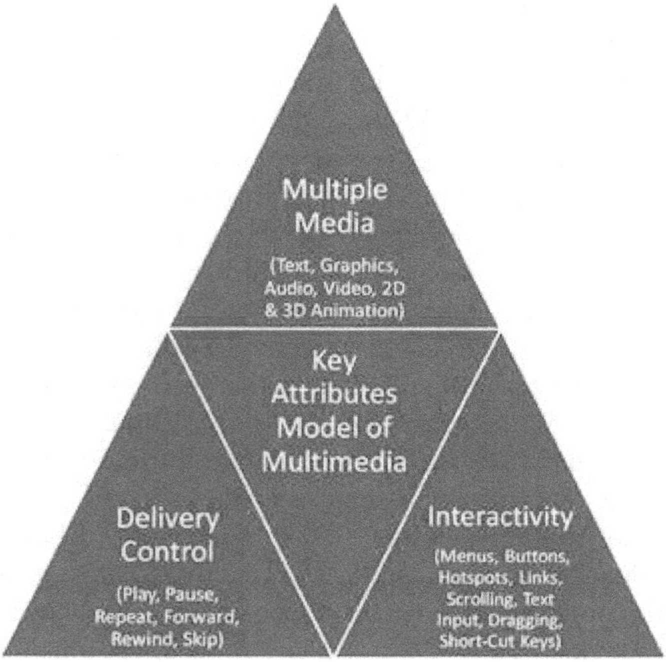

students from understanding its contents, does not use a step-by-step approach in problem solving and lacks delivery control. Some detailed descriptions of the various computer-based learning packages that were developed to support engineering education can be found in the literature (Aziz, 2011; Bailey et al., 1995; Gao et al., 2011; Gupta, 2002; Hmelo et al., 1995; Hsieh et al., 2005; John & Abdollah, 1998; Lynn et al., 1997; Mackenzie et al., 2001; Marsico, 2001; Noronha et al., 2000; Norrie, 1996; Ranga & Gramoll, 1999; Scott, 1996).

In general, the computer-aided learning packages for engineering education can be broadly divided into six main categories which are simulations, intelligent tutors, remote laboratories, interactive instructional packages, mixed realities and multi-approaches packages. Each type of the computer-aided learning packages serves different learning purposes (see Table 6). Interestingly, it is quite common to identify the redundancy in the purposes and roles played by each type of the CAL packages as illustrated in Table 6.

Table 6. Types of CAL and their purposes (Adopted from Lee & Manjit Sidhu, 2013)

Types of CAL	Purposes
Simulations	To aid in the design and visualization of engineering models in 2-D/3-D dynamics form.
	Allow the students to perform active interaction to manipulate the engineering models.
Intelligent Tutors	To guide, aid and assist the students in understanding and applying of basic engineering concepts through artificial intelligent mechanism.
	To act as the drill and practice platform for engineering students in engineering problem solving.
Remote Laboratories	To partially replace live experimentation and reduce the cost of expensive equipment (Froyd et al., 2012)
Interactive Instructional Packages	Act as the supplementary individualize learning applications for the students to learn/ revise the engineering modules through the presentation of multimedia elements (text, graphics, audio, video and animation). Usually will include the lecture, tutorial, exercises and demonstration activities in a courseware.
Mixed Reality	To create a new learning environment to support engineering visualization by integrating the virtual models into the real scenes with the support of mixed-reality technologies.
Multi-Approaches	To create a virtual learning environment for engineering education by including multiple representation approaches (simulations, intelligent tutors, coach based, cooperative learning and etc.) into an integrated learning platform.

Factors Influencing the Success of CAL in Engineering Education

The computer aided learning that used to support engineering education will not achieve its goal if technology components are considered as the sole factor involved in the CAL development. In fact, Norman (1993) claimed that technologies are rather neutral in its role and neither is inherently good or bad. It is the learning environments created that will facilitate and support the learning process. Figure 3 depicts the proposed factors that contribute to the creation of CAL environments for engineering education (Lee & Manjit Sidhu, 2013). A suitable computer aided learning environment that may facilitate the learning process should not simply just replicate efforts of the traditional learning environment with the ICT overlaid on it. Rather, it should be a completely redesign of the whole learning environment that involves at least the consideration from the instructional design perspective and the technical design perspective (Ko & Rossen, 2005). According to Renes and Strange (2011), an effective instructional design depends on two basic requirements which include the pedagogical knowledge and the content knowledge. The importance of pedagogical factors that play a significant role

Figure 3. Factors contribute to the creation of CAL environments for engineering education

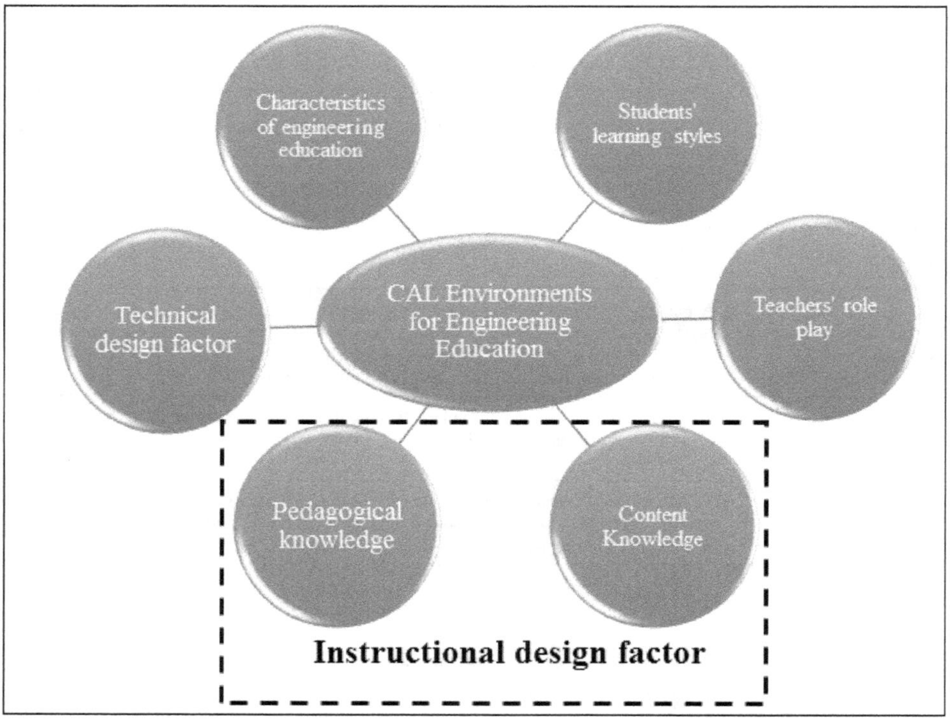

in determining the successful implementation of CAL was critically discussed in the literature (Ahern et al., 2006; Albion & Gibson, 2000; Clark, 1994; Ferdig, 2006; Park & Hannafin, 1993; Tamim et al., 2011; Zheng et al., 2008). In the context of educational psychology, there are three popular categories of learning theories, namely, behaviorism, cognitivism and constructivism. Each of these categories has its own significant characteristics that may impact on the instructional design of the learning tools that should be taken into serious consideration when designing and developing the CAL packages. Descriptions of these learning theories can be found in the literature (Atkins, 1993; Bransford et al., 2006; Fosnot, 1996).

From the technical design perspective, the user interface designs of the learning tools is a significant factor that should be taken into serious consideration when designing the CAL packages. User interface is defined as the part of computer system with which a user interacts in order to undertake his or her tasks and achieve his or her goal (Stone et al., 2005). This includes the study of screen layouts design and the display menu design (Renes & Strange, 2011). When the learners performed the study with the aid of ICT, the interfaces will act as the bridge between the computer systems and the learners. As emphasized by Liang and Sedig (2010), "*even though a tool might be pedagogically sound, but if it is not properly designed, it might not achieve the intended learning goals*" (p.989). As claimed by Schar et al. (2000), the current research on CAL is too focused on the issue related to learning content representation while neglect on the impact of user-interface in the process of learning. Proper design of the user interface for the CAL packages is important since the students (learners) will directly interact with the interface while performing the learning procedures through the use of CAL packages.

In order to have a proper computer aided learning environment, the students learning styles should not be neglected. As known, learning styles are defined as the characteristic cognitive, affective, and psychological behaviours that serve as relatively stable indicators of how learner perceives, interacts with, and responds to the learning environment (Keefe, 1991). In the process of teaching and learning, we should be aware that different students are comfortable with different learning styles (Felder & Brent, 2005). In the literature, there are four widely accepted learning style models in engineering education context (Manjit, 2006; Ogot & Okudan, 2006), which are the Myers-Briggs Type Indicator (MBTI), Kolb's Learning Style Model, Felder-Silverman Model and Honey and Mumford Learning Styles Evaluation. All these learning style instruments can be used to measure and identify the learners' preferences. A better understanding of the students learning styles may help the educator to design for better learning strategies in the CAL learning environments that may cater for different students' preferences in learning.

As commonly known, in the traditional learning environment, the teachers (instructors) play a central and significant role in the teaching and learning process. In fact, there is still insufficient definition regarding the role played by the teacher in the computer aided learning environments (Urhahne et al., 2010). As claimed by Lai (1993), the role of the teacher is to act as the knowledge facilitator instead of the knowledge presenter in a computer-supported learning environment. This claim seems to be consistent with the research findings by Zhu (2010) which found that teachers of facilitator/expert profile and facilitator/delegator profile are more inclined to adopt educational technology. In addition, Urhahne et al. (2010) further proposed a 5E model (envision the lesson, enable collaboration, encourage students, ensure learning, evaluate achievement) that illustrates the roles played by a teacher which cross the whole instructional process from lessons preparation till the assessment of achievement. However, Urhahne claimed that this model still needs further investigation on its effectiveness to improve students' achievements. In the engineering education context, with the paradigm shift in engineering education, the emphasis will be more on active learning that involves critical thinking and problem solving approach. With the shift towards computer-supported learning environment, the emerging new roles of the engineering teachers are unavoidable and need the attention urgently from the engineering educators. Intensive empirical research is urged to ensure clearer and solid definitions on the new roles played by teachers in engineering education domain and its effectiveness in integrating with the computer-supported learning environment.

Implementation of CAL Applications in Engineering Education

In engineering education, numerous studies have been carried out to implement Computer-Aided Learning (CAL) packages that incorporate multimedia and associated technologies in the educational context. A list of such multimedia CAL packages reported in the literatures is presented in Table 7.

Technology Assisted Problem Solving Packages

TAPS packages are specialized computer programs developed to work as stand-alone (PC Based) or with Web servers that can supplement student learning; for revision, laboratory experiments, and self-study founded by Manjit (2006). Although many computer-based learning environments have emerged in general as discussed in the sections titled "Computer-Aided Learning (CAL) for Engineering Education," "Overview of Learning Styles," and "Theory of Learning Styles," the term technology assisted problem solving (TAPS) environment would be used to refer to the packages that aimed at coaching students in learning particularly slow learners (students who have difficulties in understanding the concepts in Engineering mechanics dynamics subject), on the best approach to solve a particular engineering problem in a step-by-step or logical approach (Manjit, 2009). Although a TAPS package may be considered as a form of CAL, more specifically, TAPS packages were designed to allow independent problem solving, develop logical thinking, and promote learning of the subject matter. The research study on TAPS packages found

Table 7. Multimedia CAL packages developed for engineering education (Adapted from Manjit, 2006)

CAL Packages	Educational Objectives	References
Multimedia Tutorials Package (MTP)	To replace traditional lecture part of a course in engineering materials with studio exercises that could help students visualize 3-D evolving processes that cannot be presented effectively using static illustrations.	McMahon (2000)
Multimedia Based Environment (MBE)	To provide adequate MBE tutorials on engineering mechanics to motivate students in learning.	Noronha et al. (2000).
Interactive Multimedia Intelligent Tutoring System (IMITS)	To permit individual tutoring of students, i.e. allow students to solve assignments in any manner the student chooses and is able to determine the student's ability to apply and understand basic concepts of engineering.	Brain & Duarte (2002)
Amoco Computer Simulation (ACS)	To innovate teaching of engineering design by developing problem solving skills.	Mackenzie et al. (2001)
Computer Graphics Simulation (CGS)	To provide multimedia dynamic solid models for visualization.	Wilson (2001)
Multimedia Mechanics of Materials Laboratory	To familiarize students with testing equipment, data acquisition, testing procedures and reporting results.	Salvatore (2001)
Interactive Virtual Tutor (IVT)	To guide and provide helpful hints to student in an intelligent and interactive manner in solving problems in engineering mechanics statics	Gupta (2002)
Intelligent Practice Environment (IPE)	To allow the user to explore complex relationships through discovery and problem solving activities in a 3-D environment.	Janet (2002)
Interactive Tutoring Components (ITC)	To help students understand the engineering mechanics dynamics subject with the development of multiple interactive tutoring components	Shang et al. (2005)
An Interactive Multimedia E-Learning System (IMELS)	To provide students with a comprehensive problem-based learning environment for the discipline of industrial engineering.	Lau et al. (2006)
Augmented Reality and Web 3D in Engineering Education	To allow students to interact with 3D web content using virtual and augmented reality. Learners of this interface can view multimedia content, to support a lecturers traditional delivery, either locally or over the Internet and in a table-top augmented reality environment.	Liarokapis et al. (2007)
Technology Assisted Problem Solving (TAPS) Packages	The TAPS packages aimed at coaching students in learning particularly slow learners (students who have difficulties in understanding the concepts in Engineering mechanics dynamics subject) on the best approach to solve a particular engineering problem in step-by-step or logical approach.	Manjit (2006)
Interactive animation software	This interactive animation software for an introductory Thermodynamics course has been developed, which is intended to be part of the WileyPLUS platform (John Wiley and Sons, Inc.). It is unique because each animation is directly linked to a homework problem and no programming is required of the user. The animations are web-based (hard-coded in Adobe® Flash Action Script), so no external computer programs are needed.	(Stanley & DiGiuseppe, 2009)
Mobile Learning Application in Augmented Reality Environment	A mobile learning application that uses modified Reciprocal Teaching method provided as a learning aid tool for assisting student in visualizing the concept.	(Salam, 2011)
Simulation software	In this work, the researchers intended to test whether computer simulation activities can help students overcome difficulties in mechanics thought the used of teaching software (Dynamic mechanics).	(Hassouny, Kaddari, Elachqar, & Alami, 2013)
Simulation based Virtual Laboratories	Two simulation based virtual laboratories were developed to look into the empirical foundation of classical dynamics for physics learning.	(Ré & Giubergia, 2013)
Interactive Dynamics Learning Course (IDLC) web based simulations	Simulations of different mechanisms studied in the IDLC will be made available to the students on the course web site to supplement the interactive dynamics course in the classroom	(Mehendale & Irwin, 2015)

CAL Packages	Educational Objectives	References
Visualization tool using Sim2Bil	This research study utilized the visualization tool Sim2Bil, which combines a simulation of two cars, velocity graphs, and an input for velocity functions to facilitate mathematics education in engineering studies.	(Hogstad, Isabwe, &Vos, 2016)
Interactive Computer Simulation and Animation (CSA) module	A new interactive CSA module was developed to improve student learning of particle kinetics in an undergraduate engineering dynamics course. The unique feature of this CSA module is that it integrates computer visualization with mathematical modeling, so students can directly connect engineering dynamics phenomena to underlying mathematics.	(Fang & Guo, 2016)

Figure 4. Extended key attributes model of multimedia and DVR and AR

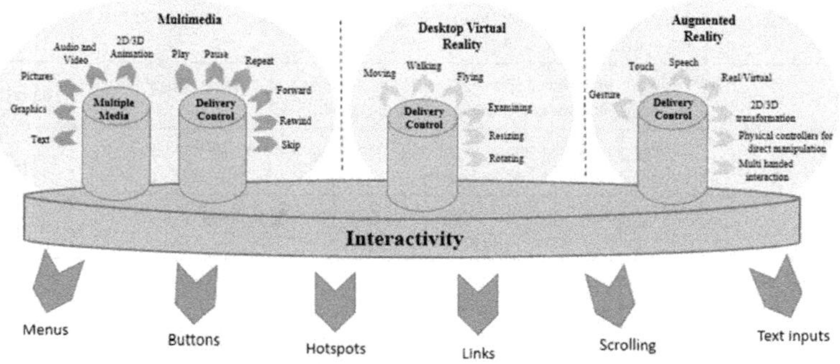

that students benefited and appreciated the most from such a TAPS package and they also found that the course becomes more interesting, enjoyable, easy and understandable when compared to the traditional method of classroom learning (Manjit, 2006). The reasons for employing TAPS packages can be summarized as follows (Manjit, 2009):

1. to use and store the knowledge of experienced instructions (human) and make the same easily accessible to the students;
2. to develop a suitable user interface for simplifying the difficult engineering concepts
3. to help slow learners acquire problem solving skills
4. to provide encouragement to students in independent learning by incorporating simple intelligent (expert system like rules) in the TAPS packages; and

5. as an attempt to improvise the limitations of the already existing computer based learning packages thereby making them more acceptable as effective learning aids in UNITEN.

As described by Manjit (2006), the TAPS packages employed a variety of multimedia elements such as text, 2-D animated and still graphics, 3-D animated and still geometric models, audio, video and animations, stereoscopic images, desktop virtual reality (DVR) and simple artificial intelligent techniques to develop individualized computer based learning environment. The extended key attributes model of multimedia and DVR and AR can be referred to Figure 4.

The extended attributes as stated in the Figure 4 are important because without it, the package cannot be fully adaptive to the individual student's on-going learning and problem solving needs during instruction. Four design approaches were designed and tested by Manjit (2006) (see Figure 5) in developing the TAPS packages namely 2-D graphics and animation (design approach 1), coach-based environment (design approach 2), 3-D virtual environment (design approach 3), and desktop virtual reality (design approach 4). These TAPS packages can be classified as cognitive tools for learning, problem solving, testing, and simulation. The summary of key features and differences of each of these four TAPS package is provided in Table 8 (Manjit, 2006).

This research aimed to improve the user interface for the TAPS package (enhancement of multimedia TAPS package). Detail descriptions on the development process for the enhanced multimedia TAPS package embedded with 25 proposed patterns of interactions are discussed in details (see Chapter 6).

LEARNING STYLES

In the era of post-capitalist knowledge society, paradigm shift in engineering education is unavoidable in response to the rapid changes in the global market environment that emphasize on the innovation efforts for competitive advantage. Many practitioners in the engineering industry look seriously into this issue and did propose for new paradigms of engineering education in response to the changes in today's increasingly knowledge driven environment. The details can be found in numerous research papers and reports (Chua, 2014; Froyd et al., 2012; Mistree et al., 2014; National Academies of Science,

Table 8. Summary of key features and differences of each TAPS and ATAPS package (Adapted from Manjit, 2006)

Key Features	2-D TAPS Package	Coach Based TAPS Package	3-D TAPS Package	3-D Desktop Virtual and Augmented Reality ATAPS package
Objective	• To examine various levels of interaction for a tutorial and problem-solving topic on Structural Analysis.	• To provide an environment that could help the student solve Engineering Mechanics problems without the help of a human instructor.	• To incorporate interactivity multimedia to study curvilinear motion of cylindrical components in a 3-D virtual environment.	• To study curvilinear motion of cylindrical components where the user could interact in a desktop virtual reality (DVR) environment for greater interaction and visualization.
Design approach	• Employed 2-D graphics/ images, colors, and tweening technique to produce animations.	• Employed 2-D graphics/ images, colors, sound, and tweening technique to produce animations. • Provide user with more dynamic and interactive support. • Provide user support tools such as calculator and notepad.	• Employed 2-D graphics/ images, 3-D geometric model, colors, sound, and tweening technique to produce animations. • Provide user with more dynamic and interactive support in a 3-D virtual environment. • Provide user support tools such as calculator and notepad. • Allow user to plot curvilinear path of a moving object in a 3-D virtual environment.	• Employed 2-D graphics/images, 3-D geometric model, colors, sound, and tweening technique to produce animations. • Provide user with more dynamic and interactive support in a 3-D virtual environment. • Provide user support tools such as calculator and notepad. • Allow user to plot curvilinear path of a moving object in a 3-D virtual environment and real time interaction. • Provides animated and static stereoscopic images. • Virtual objects should interact with the user and real objects in the scene in a natural manner.
Learning/ Problem solving approach	• Provides learning objectives to user. • A sequence of steps and solution of the problem are presented to the student. The student moves forward to the next step or back to the previous step or solution.	• Provides learning objectives to user. • A sequence of steps and solution of the problem are presented to the student. The student moves forward to the next step or back to the previous step or solution. • Allows user to select a formula from a list provided and validates if the correct formula has been applied to solve a problem. • Allows user to make mistakes. • Employed simple expert system rules to coach user.	• Provides learning objectives to user. • A sequence of steps and solution of the problem are presented to the student. The student moves forward to the next step or back to the previous step or solution. • Allows text, numeric and special characters as input data.	• Provides learning objectives to user. • A sequence of steps and solution of the problem are presented to the student. The student moves forward to the next step or back to the previous step or solution. • Allows text, numeric and special characters as input data. • Employed simple expert system rules to coach user.
Association	• The user needs to be able to associate the virtual environment of 2-D elements and hid/her interactions with related information such as images, textual resources, and the other data with the problem solving process performed by the user.	• The user needs to be able to associate the virtual environment of 2-D elements and hid/her interactions with related information such as images, textual resources, and the other data with the problem solving process performed by the user.	• The user needs to be able to associate the virtual environment of 3-D geometric elements and hid/her interactions with related information such as images, textual resources, and the other data with the problem solving process performed by the user.	The user needs to be able to associate the virtual environment of 3-D geometric elements and hid/her interactions with related information such as images, textual resources, and the other data with the problem solving process performed by the user.

continued on following page

Table 8. Continued

Key Features	2-D TAPS Package	Coach Based TAPS Package	3-D TAPS Package	3-D Desktop Virtual and Augmented Reality ATAPS package
Coaching	• Coaching is provided to enhance the user's problem solving experience while performing complex tasks in 2-D environment.	• Coaching is provided to enhance the user's problem solving experience while performing complex tasks in 2-D environment. • The built-in expert system rules manage the user's activities in the 2-D problem solving environment and provides dynamic coaching advice and feedback based on the user's activities in a 2-D coach based environment.	• Coaching is provided to enhance the user's problem solving experience while performing complex tasks in 3-D environment.	• Coaching is provided to enhance the user's problem solving experience while performing complex tasks in 3-D environment. • The built-in expert system rules manage the user's activities in the 3-D problem solving environment and provides dynamic coaching advice and feedback based on the user's activities in a 3-D coach based environment.
Feedback and Assessment	• For a positive experience, the user is given feedback and assessment to understand his/her progress.	• For a positive experience, the user is given feedback and assessment to understand his/her progress • Allow text, numeric and special characters as input data. • User' progress i.e. scores is stored in database. • The problem solving environment provides timely feedback and assessment directly related to the user's interactions. The feedback helps increase the user's ability to reason and analyze the problem solving environment.	• For a positive experience, the user is given feedback and assessment to understand his/her progress	• For a positive experience, the user is given feedback and assessment to understand his/her progress • Allow text, numeric and special characters as input data. • Multiple choice questions and answers are randomized. • The problem solving environment provides timely feedback and assessment directly related to the user's interactions. The feedback helps increase the user's ability to reason and analyze the problem solving environment.
Text/ Contents	• Minimum text of theory is used in order to make use of multimedia elements.	• Minimum text of theory is used in order to make use of multimedia elements.	• Minimum text of theory is used in order to make use of multimedia elements.	• Minimum text of theory is used in order to make use of multimedia elements.

continued on following page

Engineering, and Medicine, 2007; National Science Board, 2007; Prados, 1998; Rajala, 2012; Rosen, 2007; Wince-Smith, 2005). In order to further enhance the quality of teaching and learning in engineering education, the students' preference in learning is an important factor that should raise the attention of the education practitioners. This referred to the learning styles of the students. Learning styles are defined as "the characteristic cognitive, affective,

Table 8. Continued

Key Features	2-D TAPS Package	Coach Based TAPS Package	3-D TAPS Package	3-D Desktop Virtual and Augmented Reality ATAPS package
Interactivity	• The student interacts and observes meaningful tasks i.e. such as movement of the support and rotation of structure.	• The student interacts and observes meaningful tasks i.e. such as movement of the support and rotation of structure. • Provides the user with more dynamic and interactive support than what has been provided in existing computer aided learning packages such as motion, and feedback response in the event when student makes mistake. • Uses reasoning support (simple expert system rules are used) and explanations of complicated concepts while the user is trying to solve the problem presented in the package. Thus this approach can help students make decisions and complete tasks better and also provide explanation for reasoning, enabling continual performance and improvement.	• The student interacts and observes meaningful tasks i.e. such as curvilinear motion of cylindrical components in a 3-D virtual environment. • A simple algorithm that allows student to plot path / path in a 3-D virtual environment.	• The student interacts and observes meaningful tasks i.e. such as movement of the support and rotation of structure. • A simple algorithm that allows student to plot path in a 3-D virtual environment. • Gesture, touch, speech • Physical controllers for direct manipulation • Multi handed interaction

continued on following page

and psychological behaviors that serve as relatively stable indicators of how learner perceives, interacts with, and responds to the learning environment" (Keefe, 1991, p. 4). The engineering instructors, should be aware that different students are comfortable with different learning styles (Felder & Brent, 2005). The understanding of students' preferences may contribute to the adjustment of teaching strategies and the design of learning instructions that will better accommodate for students learning needs (Cavanagh & Coffin, 1994; Chen & Chiou, 2012; Graf, 2007; Graf et al., 2007; Noguera & Wageman, 2011; Pedrosa de Jesus et al., 2004).

Table 8. Continued

Key Features	2-D TAPS Package	Coach Based TAPS Package	3-D TAPS Package	3-D Desktop Virtual and Augmented Reality ATAPS package
General Outcomes and Benefits	• The use of tweening technique was found to be helpful where animations were used to illustrate motions such as movement of the support and rotation of structure. • The use of 2-D animated graphics and colors helped students understand better in solving the problem presented in the TAPS package. • Benefits noted from this package include interactive multimedia provide comprehensive coverage combining full motion, audio, video and 2-D animated graphics. • Interactive multimedia provides the student as integrated learning environment, which combines explanations with illustrative examples, this is in agreement as stated by Cairncross (2002).	• The TAPS package help students to manage the sequence of steps the student should perform to solve the problem and control the 2-D animated mechanisms leading from problem statement till the solution. • It is generally accepted that this TAPS package can help students process better cognitive-perceptions which can result in fast and better understanding of the problem. • Benefits noted from this package include interactive multimedia provide comprehensive coverage combining full-motion, audio, video and 2-D animated graphics.	• Helped students to understand and visualized the problem on curvilinear motion which, was otherwise difficult to explain to students. • Dynamic models present challenges to students beyond what they have learnt in the traditional way, this is in agreement as stated by Liang (2002). • Benefits noted from this package include interactive multimedia provide comprehensive coverage combining full-motion, audio, video, 2-D and 3-D animated graphics. • More emphasis is given to the visualization of 3-D objects because 3-D could enhance the process of leaning, increase the level of understanding enabling students to understand more effectively through interactivity. This environment can provide a rewarding learning experience that would be otherwise difficult to obtain (Liarokapis et al. 2007)	• Allow student to experiment simulated problem-solving problem in 3-D environment and manage a complex task in real time. • Stereoscopic images and animated video files gives a better 3-D view of the robotic arm and path as it is not the same as static image as shown in the textbooks. The stereoscopic images help students enhance depth perception that could reduce learning time as compared to the conventional. • It is generally accepted that the stereoscopic images could create great interest and enthusiasm for students in understanding the problem presented in the TAPS package, particularly, students experiencing difficulties in understanding the Engineering Mechanics Dynamics subject. • Multimedia and virtual reality techniques with simulation initiates a new appearance for learning applications –real time presentations of 3-D data, this is in agreement as stated by Klett (2002). • With simulations, students are able to take a more active role in learning, this is in agreement as stated by Jesica and Tara (2005). • Dynamic representations enable more efficient communication of complex concepts; this is in agreement as stated by Hennessy et al. (2007). • Blended real and simulated environment.

Overview of Learning Styles

The research on learning styles has been active since four decades ago (Cassidy, 2004; Pedrosa de Jesus et al., 2004). There exist various definitions for learning styles. According to Campbell et al. (2003), learning styles is defined as a certain specified pattern of behavior according to which the individual approaches learning experience. While Felder and Spurlin (2005)

Figure 5. Four main design approaches used for ATAPS packages

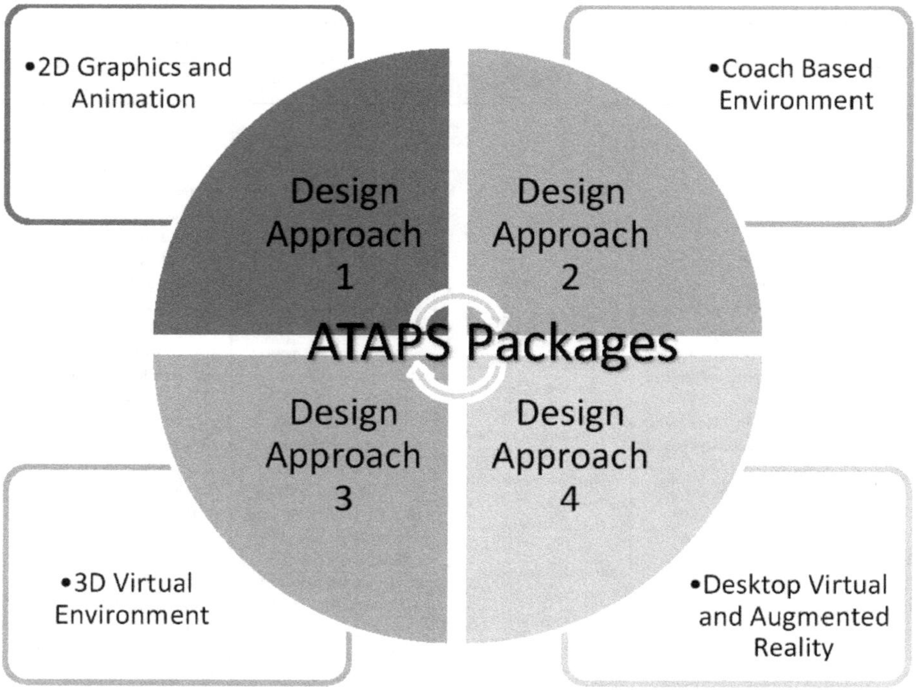

Figure 6. Learning styles and learning cycle based on Kolb's Model (Adopted from Montgomery & Groat, 1998)

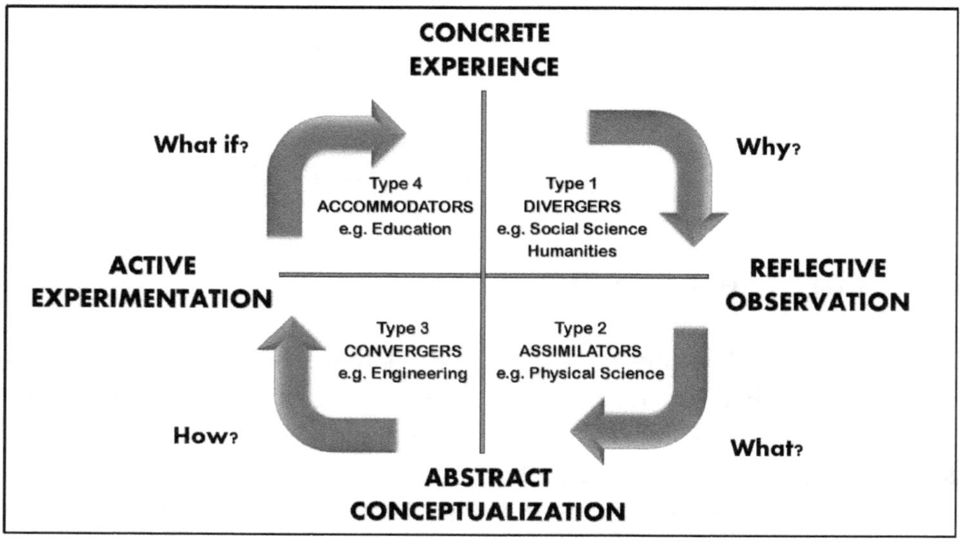

Table 9. The Honey and Mumford's four learning styles (Honey & Mumford, 1986)

Learning styles	Strengths	Weakness
Activist (learn best when they are actively involve in new tasks)	Sociable, open-minded, welcome challenge, highly involved, prefers here-and-now	Bored by implementation details and the longer term, always seeks the limelight
Reflector (learn best through review and reflect)	Good listener, tolerant, sees different perspectives, postpones judgement, cautious	Takes a back seat in meetings, low profile, distant
Theorist (learn best when they can relate new information to concept & theory)	Integrates observations with theory, rationale, objective, analytical	Perfectionist, detached, impatient with subjective and intuitive thinking
Pragmatist (learn best when they see relevance of real life issues)	Experimenter, quick to adopt and try out new ideas, practical, down-to-earth	Impatient with theory, impatient with open-ended discussion

defined learning styles as the different ways students take in and process information. Another popular definition for learning styles refers to individuals' characteristics and preferred ways of gathering, organizing and thinking about information (Fleming, 2005). As noted by Kolb (1983), learning styles are not fixed personality traits but rather one's adaptive orientation to learning. Felder and Spurlin (2005) shared similar view with Kolb by which they stressed that "*learning style profiles suggest behavioral tendencies rather than being infallible predictors of behavior*" (p. 104). Many of the researchers did agree that individuals may tend to have a preference for one or two learning styles over others and the preferences can be affected by a student's educational experience (Felder & Spurlin, 2005; Honey & Mumford, 1992; Kolb, 1983). Therefore, learning styles are "*relatively stable but are not immutable*" (Pedrosa de Jesus et al., 2004, p. 533). Throughout the learning process and based on different educational experience, the students may discover better way of learning and develop certain learning preferences. The learning style assessments can benefit both the instructors and students (Felder & Spurlin, 2005; Larkin-Hein & Budny, 2001). Thus, from the instructors' perspective, the identification of the students learning styles may lead to gain better understanding of learners with different learning styles. This may contribute to assist the design of teaching and learning instructions to accommodate the different learning needs of students. While from the students' perspective, better understanding of the learning styles may provide ideas on how the students might further improve on the less preferred styles and overcome the learning difficulties by developing the skills for balance approach in effective learning.

Table 10. The four dimensions of Felder and Silverman's learning styles

Learning Styles	Descriptions
Sensory/Intuitive	Sensors prefer facts, data, experimentation, sights and sounds, physical sensations are careful and patient with details, but may be slow. Intuitions prefer concepts, principles and theories, memories, thoughts, insights and may be quick but careless.
Visual/Verbal	Visual learners prefer pictures, diagrams, charts, movies, demonstrations and exhibitions. Verbal learners prefer words, discussions, explanations, written and spoken explanations, formulas and equations.
Active/Reflective	Active learners learn by doing and participating through engagement in physical activity or discussion. Reflective learners learn by thinking or pondering through introspection.
Sequential/Global	Sequential learners take things logically step by step and will be partially effective with understanding. Global learners must see the whole picture for any of it to make sense and are completely ineffective until they suddenly understand the entire subject.

There are considerable amount of literature regarding the issue of learning styles (Cassidy, 2004; Coffield et al., 2004a; Coffield et al., 2004b; Deborah et al., 2012; Kolb et al., 2001; Riding & Rayner, 1998; Sadler-Smith, 1997; Sternberg &Grigorenko, 1997) and various instruments for learning styles measurements throughout all these years (Cornwell &Manfredo, 1994; Dunn & Dunn, 1979; Felder & Silverman, 1988; Fleming, 2005; Gregorc, 1985; Honey & Mumford, 1992; Kolb, 1983; McCaulley, 2000). As argued by Hawk and Shah (2007), although various different instruments are available for learning styles measurement, the learning styles instruments or inventories vary in length, format and complexity, and no single instrument can capture all of the richness of the phenomena of learning styles.

Recently, engineering educators have been increasingly taking the learning style theories into serious consideration for classrooms teaching and learning (for example, Cagiltay, 2008; Felder, 1996; Felder & Brent, 2005; Felder & Silverman, 1988; Holvikivi, 2007; Manjit, 2006; Lee & Manjit Sidhu, 2013a; Miskioglu & Wood, 2013; Rosati et al., 1988). In the literature, there are four widely accepted learning style models in engineering education context, which are the Myers-Briggs Type Indicator (MBTI), Kolb's Learning Style Model, Felder-Silverman Model and Honey and Mumford Learning Styles Questionnaires (Manjit Sidhu, 2006; Ogot & Okudan, 2007). All these learning style instruments can be used to measure and identify the learners' preferences.

Table 11. Reported learning preferences (adapted from Felder and Spurlin, 2005)

Sampled population	A	S	Vs	Sq	*N*	Reference
Iowa State, Materials Engr.	63%	67%	85%	58%	129	Constant (1997)
Michigan Tech, Env. Engr.	56%	63%	74%	53%	83	Paterson (1999)
Oxford Brookes Univ., Business	64%	70%	68%	64%	63	Vita (2001)
British students	85%	86%	52%	76%	21	
International students	52%	62%	76%	52%	42	
Ryerson Univ., Elec. Engr.						
Students (2000)	53%	66%	86%	72%	87	Zywno&Waalen (2001)
Students (2001)	60%	66%	89%	59%	119	Zywno (2002)
Students (2002)	63%	63%	89%	58%	132	Zywno (2003)
Faculty	38%	42%	94%	35%	48	
Tulane, Engr.						
Second-Year Students	62%	60%	88%	48%	245	Livesay et al. (2002)
First-Year Students	56%	46%	83%	56%	192	Dee et al. (2003)
Universities in Belo Horizonte (Brazil)						
Sciences	65%	81%	79%	67%	214	Lopez (2002)
Humanities	52%	62%	39%	62%	235	
Univ. of Limerick, Mfg. Engr.	70%	78%	91%	58%	167	Seery et al. (2003)
Univ. of Michigan, Chem. Engr.	67%	57%	69%	71%	143	Montgomery (1995)
Univ. of Puerto Rico-Mayaguez						
Biology (Semester 1)	65%	77%	74%	83%	39	Buxeda& Moore (1999)
Biology (Semester 2)	51%	69%	66%	85%	37	Buxeda& Moore (1999)
Biology (Semester 3)	56%	78%	77%	74%	32	Buxeda& Moore (1999)
Elect. & Comp. Engr.	47%	61%	82%	67%	?	Buxeda et al. (2001)
Univ. of Sao Paulo, Engr.	60%	74%	79%	50%	351	Kuri&Truzzi (2002)
Civil Engr.	69%	86%	76%	54%	110	
Elec. Engr.	57%	68%	80%	51%	91	
Mech. Engr.	53%	67%	84%	45%	94	
Indust. Engr.	66%	70%	73%	50%	56	
Univ. of Technology Kingston, Jamaica	55%	60%	70%	55%	?	Smith et al. (2002)
Univ. of Western Ontario, Engr.	69%	59%	80%	67%	858	Rosati (1999)
First year engr.	66%	59%	78%	69%	499	Rosati (1996)
Fourth year engr.	72%	58%	81%	63%	359	Rosati (1996)
Engr. Faculty	51%	40%	94%	53%	53	Rosati (1996)
Brunel Univ. UK. IS & Computing	64%	70%	82%	68%	148	Baldwin &Sabry (2003)
Universidad de las Americas, Puebla	67%	82%	90%	55%	290	Palou (2006)
California Polytechnic State Univ.	56%	74%	79%	71%	86	Self &Widmann (2009)

continued on following page

Table 8. Continued

Sampled population	A	S	Vs	Sq	*N*	Reference
Auckland Univ., Chemical & Materials Engr. course	34%	69%	81%	72%	29	Patterson (2011)
Utah State Univ., Mechanical & Aerospace Engr. course	42%	89%	89%	69%	61	Fang & Zhao (2013)
Beijing ForestryUniv., Civil Engr.	55%	80%	76%	55%	71	Fang & Zhao (2013)
The HK Institute of Education	56%	66%	81%	59%	32	Cheng (2014)
Texas Tech University	51%	69%	82%	65%	51	Hames & Baker (2015)
Durban University of Technology (Industrial Engineering)					200	Jackson (2015)
Level 1 Engineering Students	55%	66%	61%	57%	?	
Level 2 Engineering Students	56%	65%	62%	57%	?	
Level 3 Engineering Students	57%	65%	69%	55%	?	
United Arab Emirates University	57%	71%	83%	68%	118	Chowdhury (2015)
National Institute of Technical Teachers Training & Research, Chandigarh.						PK et al. (2016)
CS & Engineering	69%	69%	91%	75%	32	
Civil Engineering	50%	70%	87%	70%	30	
Electrical Engineering	62%	57%	95%	54%	37	
Electronic & Communication Engineering	55%	65%	90%	65%	31	
Mechanical Engineering	49%	60%	96%	67%	45	

Theory of Learning Styles

Kolb's Experiential Learning Model

Kolb developed the learning style inventory (LSI) in 1976 and revised in 1985 (Tendy & Geiser, 1997). In Kolb's model, students are classified as having a preference for (a) concrete experience or abstract conceptualization (how they take information in) and (b) active experimentation or reflective observation (how they process information) (Cornwell & Manfredo, 1994; Kolb, 1983; Stice, 1987). Figure 6 illustrates the learning styles and learning cycle based on Kolb's model.

The four types of learners in this classification (Figure 6) scheme are:

Table 12. Detail descriptions of three key areas of learning preferences

Areas	Preference style	
	Structured	Spontaneous
Approach to learning (To what extent the learners need structure and organization during learning?)	Structured learners are more likely to: • Like well-organized environments and therefore might feel uncomfortable in more ambiguous situations. • Prefer their learning to be well structured and formally planned. • Are good at following step-by-step procedures. • Be more likely to maintain focus and avoid distractions. • Prefer to stick with tried-and-tested approaches and methods. • Get uncomfortable if things are left to "loose" or they do not know what is coming up.	Spontaneous learners are more likely to: • Learn through trial-and-error • Be content with lecturers to give them "loose" and brief guidelines and they will be happy to get on with their assignments • Be happy with less structured approaches to learning. • Need lots of variety in their day and may get bored with routine. • Get stuck in; can be impatient with instructions or briefings • Enjoy spontaneity and are not worried that they don't know what is coming up next.
	Pragmatic	Conceptual
Focus on learning (How interested are the learners in the underlying concepts and workings?)	Pragmatic learners are more likely to: • Focus on practical aspects e.g. how useful, and how can they apply the learnt skills to something? • Tend to be focused on concrete, more immediate benefits of learning. • Don't necessarily see the point of being absorbed into the theory or spending time on conceptual discussions. • Believe in keeping things neat and simple. • Make their minds up quickly, think on their feet. • Prefer hands-on practical jobs, perhaps more suited to vocational education • Be more focused with the task at hand and not likely to be side-tracked by conceptual details. • Be seen as "down-to-earth," having "common sense" and good at getting things done	Conceptual learners are more likely to: • Enjoy understanding how things work from a theoretical perspective • They are more likely to enjoy complex, theoretical thinking about subjects such as Psychology. • Spend time thinking about concepts and taking it to a deeper level of understanding, perhaps to appreciate wider possibilities and related subjects or information. • Appreciate the logic and rationale behind proposed procedure • Be more curious about how the world around them "works," more likely to ask "why?" or "how?" in their mind. • Carefully weigh things up and therefore on occasions may find it harder to be decisive about complex matters, less likely to take a clear stances towards matters. • Get occasionally engrossed in the things that personally interest them and may be lose sight of the practical goals or tasks at hand. • Very high scorers can seem to have their "head in the clouds" at times by more pragmatic people.
	Concrete	Fluid
Transfer of learning (Do the learners focus on a specific problem, or transfer learning across situations?)	Concrete learners are more likely to: • Take their learning literally i.e., this learnt skill is used for this specific situation. • Find it more difficult to adapt what they have learnt to other similar situations. • Prefer following clear instructions and to be offered or given solutions. • Need to concentrate on one thing at a time, working through information in a step-by-step fashion. • Have less need to review and explore what they can do with what they have learnt.	Fluid learners are more likely to: • Enjoy tackling several things at the same time • Like exploring the links or connections between things • Automatically consider widening the original application after learning "what else can I do with this knowledge?" • See if it is possible to transfer and adapt learning from one situation to other very different situation. • Boost their learning by drawing on their own previous experiences, perhaps from totally different areas. • Be more able to adapt to changing situations.

Figure 7. Snapshot of "Hemispheric Map" diagram indicated a particular student's learning preference and the preference side of brain when processing information

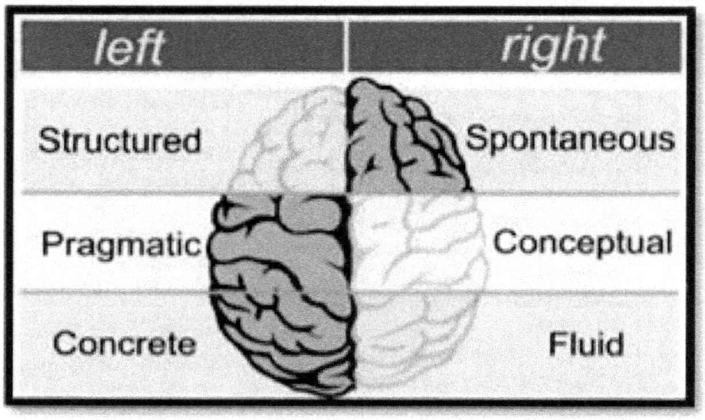

Table 13. Characteristics both for left hemisphere preference and right hemisphere preference

Left hemisphere preference	Right hemisphere preference
Likes written information, responds well to spoken instructions.	Likes to see things in action, and responds better to demonstrations (prefer demonstrations).
Solves problems in a step-by-step, logical manner.	Solves problems by using "gut feel" and hunches
Looks for differences and when things don't "fit."	Looks for patterns and links
Plans and structures.	Looks for relationships and similarities
Prefers established, more objective types of information e.g., science.	Prefers more subjective, diffuse or elusive information, e.g., art, politics.
Likes asking questions, likes things with a clear answer or predictable outcome	Happy with ambiguity, explores matters without need for a clear answer.
Less open or valuing of feelings	Responds more to feelings and open with emotions.
Splits things up - makes distinctions	Jumps in, makes things up as go along, chunks things together
Looks for cause and effect	Interested more in expression than cause and effect.

- Type 1 (concrete, reflective) – the **diverger**. Type 1 learners respond well to explanations of how course material relates to their experience, interests, and future careers. Their characteristic question is "Why?"
- Type 2 (abstract, reflective) – the **assimilator**. Type 2 learners respond to information presented in an organized, logical fashion and benefit

if they are given time for reflection. Their characteristic question is "What?"

- Type 3 (abstract, active) – the **converger**. Type 3 learners respond to having opportunities to work actively on well-defined tasks and to learn by trial-and-error in an environment that allows them to fail safely. Their characteristic question is "How?"
- Type 4 (concrete, active) – the **accommodator**. Type 4 learners like applying course materials in new situation to solve real problems. Their characteristic question is "What if?"

Honey and Mumford's Learning Styles Questionnaire (LSQ)

Honey and Mumford defined learning styles as "*a description of the attitudes and behaviours that determines an individual's preferred way of learning*" (Honey & Mumford, 2000, p. 6). Honey and Mumford proposed a model similar to Kolb's in which individuals have a mixture of four learning styles, normally with a preference for one or two of the styles. Honey and Mumford's learning styles questionnaire (LSQ) and Kolb's learning styles inventory (LSI) are both diagnostic tests to help individuals identify their strengths, weaknesses, and development needs. The four learning styles, based on Kolb's theory of stages in the learning cycle, are: activist, reflector, theorist, and pragmatist as shown in Table 9. Through the use of the LSQ instrument, instructors can gain a better understanding of individual learners' attitudes, behaviors and learning processes (Armstrong et al., 2005; Jackson, 2002).

The Myers-Briggs Type Indicator

People (genera; public) are classified on the Myers-Briggs Type Indicator (MBTI) according to their preferences on four scales derived from Jung's Theory of Psychological Types (McCaulley, 2000; Pittenger, 2005).

- **Extraverts** (try things out, focus on the outer world of people) or Introverts (think things through, focus on the inner world of ideas).
- **Sensors** (practical, detailed-oriented, focus on facts and procedures) or Intuitors (imaginative, concept-oriented, focus on meanings and possibilities).

- **Thinkers** (skeptical, tend to make decisions based on logic and rules) or Feelers (appreciative, tend to make decisions based on personal and humanistic considerations).
- **Judgers** (set and follow agendas, seek closure even with incomplete data) or Perceivers (adapt to changing circumstances, postpone reaching closure to obtain more data).

As discussed by Felder and Brent (2005), most engineering instruction is oriented toward introverts (lecturing and individual assignments rather than active class involvement and cooperative learning), intuitors (emphasis on science and math fundamentals rather than engineering applications and operations), thinkers (emphasis on objective analysis rather than interpersonal considerations in decision-making), and judgers (emphasis on following the syllabus and meeting assignment deadlines rather than on exploration of ideas and creative problem solving).

Felder-Silverman Index of Learning Styles (ILS)

The Felder-Silverman Learning Style Model classifies students along four dimensions: sensing/intuitive, visual/verbal, active/reflective and sequential/global as shown in Table 10 (Felder, 1993; Felder & Silverman, 1988). Previous studies (mainly undergraduate students in engineering domain) that adopted the Felder-Silverman ILS are summarized as shown in Table 11.

Ogden's Personality and Learning Styles Questionnaire

Personality and Learning Styles Questionnaire is a set of instrument that measures individual's learning styles. The questionnaire has been designed by Richard Ogden in UK, a professional psychologist based on reliable and valid psychometric principles (Ogden, 2007). According to Richard, it is primarily intended for those in an academic environment, although the content will be useful for anyone interested in understanding more about learning and their own, special personality preferences.

The learning style model explores three key areas, highlighting how the student may prefer to go about learning things or approaching tasks based on (i) approach to learning (Structured or Spontaneous), (ii) focus on learning (Pragmatic or Conceptual),(iii) transfer of learning (Concrete or Fluid). The detail descriptions for the three key areas of learning preferences are listed in Table 12.

As per Ogden (2007), the student learning preference can also be represented using "Hemispheric Map" Diagram (see Figure 7). It is a simpler way of representing where the students preferences are, and also shows the student which side of brain the student may prefer to use when processing information. The left hemisphere preference and right hemisphere preference both shown different and contrast characteristics as stated in Table 13.

Each individual learner may be able to identify their learning preference towards using the left side or the right side of the brain based on the report generated after the questionnaire is submitted through the online form. The report would provide specific advice to help the learners to balance his/her approach in learning and to learn more effectively.

SUMMARY

In summary this Chapter discussed the issues on engineering education and its challenges in the knowledge based society. Computer aided learning specifically in the context of engineering education and the importance of learning styles were also discussed. In this research study, the Honey and Mumford's Learning Styles Questionnaire (LSQ) and the Ogden's personality and learning styles questionnaire were identified and adopted to evaluate the engineering students learning preferences due to their good reliabilities as suggested in the literature.

REFERENCES

10.1111/j.1467-8535.2008.00846.x

ABET. (2013). *Criteria for accrediting engineering programs 2014-2015*. Author.

Achikalin, M., & Dur, E. (2005). The use of computer technologies in the social studies classroom. *The Turkish Online Journal of Educational Technology*, *4*(2), 18–26.

Aglan, H. A., & Ali, S. F. (1996). Hands-On Experiences: An Integral Part of Engineering Curriculum Reform. *Journal of Engineering Education, 85*(4), 327–330. doi:10.1002/j.2168-9830.1996.tb00252.x

Ahern, T., Shaw, S., & Liu, X. (2006). Teaching courses online: A review of the research. *Review of Educational Research, 76*(1), 93–135. doi:10.3102/00346543076001093

Albion, P. R., & Gibson, I. W. (2000). Problem based learning as a multimedia design framework in teacher education. *Journal of Technology and Teacher Education, 8*(4), 315–326.

American Society for Engineering Education (ASEE). (1987). *A national action agenda for engineering education*. Washington, DC: ASEE.

Atkins, M. J. (1993). Theories of learning and multimedia applications: An overview. *Research Papers in Education, 8*(2), 251–271. doi:10.1080/0267152930080207

Augustine, N. (2005). *National academies committee on prospering in the global economy of the 21st century: Rising above the gathering storm: Energizing and employing America for a brighter economic future*. Washington, DC: National Academies Press.

Aziz, E.-S. (2011). Teaching and learning enhancement in undergraduate machine dynamics. *Computer Applications in Engineering Education, 19*(2), 244–255. doi:10.1002/cae.20306

Bailey, J. D., Hall, J. L., Reed, P. A. S., & Colbourn, C. J. (1995). CBL in engineering: Student's use of a learning resource on phase diagrams. *Computers in Education Journal, 25*(1-2), 75–80. doi:10.1016/0360-1315(95)00043-7

Balamuralithara, B., & Woods, P. C. (2009). Virtual laboratories in engineering education: The simulation lab and remote lab. *Computer Applications in Engineering Education, 17*(1), 108–118. doi:10.1002/cae.20186

Baldwin, L., & Sabry, K. (2003). Learning styles for interactive learning systems. *Innovations in Education and Teaching International, 40*(4), 325–340. doi:10.1080/1470329032000128369

Barroso, L. R., Morgan, J., & Simpson, N. (2008). Classroom demonstrations with multiple modes: Virtual + reality = enhanced learning. In 2008 38th Annual Frontiers in Education Conference (p. S1C–7–S1C–12). IEEE. 10.1109/FIE.2008.4720566

Biggs, J., & Tang, C. (2003). *Teaching for quality learning at university: What the student does* (2nd ed.). New York, NY: Open University Press.

Bloom, B. S. (1969). *Taxonomy of Educational Objectives: The Classification of Educational Goals (Handbook I).* Longman Group.

Bonwell, C. C. (1991). *Active Learning: Creating Excitement in the Classroom.* Retrieved from https://www.ydae.purdue.edu/lct/HBCU/

Borsook, T. K., & Higginbotham-Wheat, N. (1991). Interactivity: What is it and what can it do for computer-based instruction? *Educational Technology, 31*(5), 11–17.

Bransford, J., Vye, N., Stevens, R., Kuhl, P., Schwartz, D., & Bell, P. (2006). Learning theories and education: Toward a decade of synergy. In P. Alexander & P. Winne (Eds.), *Handbook of educational psychology* (pp. 209–244). Mahwah, NJ: Lawrence Erlbaum.

Butz & Duarte. (2002). Interactive Multimedia Intelligent Tutoring System (IMITS). *Proceedings of the* 2002 *American Society for Engineering Education Annual Conference and Exposition.*

Buxeda, R. J., Jimenez, L., & Morell, L. (2001). Transforming an engineering course to enhance student learning. *Proc. 2001 International Conference on Engineering Education.*

Buxeda, R. J., & Moore, D. A. (1999). Using Learning-Styles Data to Design a Microbiology Course. *Journal of College Science Teaching, 29*(3), 159–164.

Cagiltay, N. E. (2008). Using learning styles theory in engineering education. *European Journal of Engineering Education*, *33*(4), 415–424. doi:10.1080/03043790802253541

Cairncross, S., & Mannion, M. (1999). How multimedia functions in engineering. *Engineering Science and Education Journal*, *8*(3), 100–107. doi:10.1049/esej:19990302

Campbell, L. C., Campbell, B., & Dickinson, D. (2003). *Teaching and learning through multiple intelligences* (3rd ed.). Needham Heights, MA: Allyn and Bacon.

Cassidy, S. (2004). Learning styles: An Overview of theories, models, and measures. *Educational Psychology*, *24*(4), 419–444. doi:10.1080/0144341042000228834

Cavanagh, S. J., & Coffin, D. A. (1994). Matching instructional preference and teaching styles: A Review of the literature. *Nurse Education Today, 14*(2), 106–110. doi:10.1016/0260-6917(94)90112-0 PMID:8177180

Chen, B. H., & Chiou, H. H. (2012). Learning style, sense of community and learning effectiveness in hybrid learning environment. *Interactive Learning Environments, 22*(4), 485–496. doi:10.1080/10494820.2012.680971

Cheshier, S. R. (1998). *Studying Engineering Technology.* Los Angeles, CA: Discovery Press.

Chua, K. J. (2014). A comparative study on first-time and experienced project-based learning students in an engineering design module. *European Journal of Engineering Education, 39*(5), 556–572. doi:10.1080/03043797.2014.895704

Clark, R. E. (1994). Media will never influence learning. *Educational Technology Research and Development, 42*(2), 21–29. doi:10.1007/BF02299088

Coffield, F., Moseley, D., Hall, E., & Ecclestone, K. (2004a). *Learning styles and pedagogy in post-16 learning: A Systematic and critical review.* London, UK: Learning and Skills Research Council.

Coffield, F., Moseley, D., Hall, E., & Ecclestone, K. (2004b). *Should we be using learning styles?What research has to say to practice.* London, UK: Learning and Skills Research Council.

Connelly, F. M., & Xu, S. (2012). Curriculum and curriculum studies. In J. Arthur & A. Peterson (Eds.), *The Routledge companion to education* (pp. 115–124). Oxon, UK: Routledge.

Constant, K. P. (1997). Using multimedia techniques to address diverse learning styles in materials education. *Journal Material Education, 19*, 1–8.

Cornwell, J. M., & Manfredo, P. A. (1994). *Kolb'S Learning Style Theory Revisited.* Academic Press.

Cornwell, J. M., & Manfredo, P. A. (1994). Kolb'S Learning Style Theory Revisited. *Educational and Psychological Measurement, 54*(2), 317–327. doi:10.1177/0013164494054002006

Dale, E. (1969). *Audiovisual Methods in Teaching (3rd ed.).* New York: The Dryden Press.

Deborah, L. J., Baskaran, R., & Kannan, A. (2012). Learning styles assessment and theoretical origin in an e-learning scenario: A Survey. *Artificial Intelligence Review*, *42*(4), 801–819. doi:10.100710462-012-9344-0

Dee, K. C., Livesay, G. A., & Nauman, E. A. (2003). Learning styles of first- and second-year engineering students. In *Proceedings 2003 ASEE/WFEO International Colloquium*. Nashville, TN: ASEE.

Deniz, H., & Cakir, H. (2006). Design principles for computer-assisted instruction in histology education: An exploratory study. *Journal of Science Education and Technology*, *15*(5-6), 399–408. doi:10.100710956-006-9031-5

Dhandabani, L., & Sukumaran, R. (2014). Use of ICT in engineering education: A survey report. In *2014 IEEE International Conference on Computational Intelligence and Computing Research* (pp. 1–5). IEEE. 10.1109/ICCIC.2014.7238362

Duderstadt, J. J. (2008). *Engineering for a changing world – A Roadmap to the future of engineering practice, research, and education*. Ann Arbor, MI: The University of Michigan.

Dunn, R., & Dunn, K. J. (1979). Learning Styles / Teaching Styles: Should they, can they, be matched? *Educational Leadership*, *36*, 238–244.

El Hassouny, E. H., Kaddari, F., Elachqar, A., & Alami, A. (2013). Teaching / Learning mechanics in high school with the help of Dynamic software. In *5th World Conference on Educational Sciences - WCES 2013* (pp. 4617–4621). Academic Press.

Entwistle, N., Thomson, S., & Tait, H. (1992). *Guidelines for promoting effective learning in higher education. University of Edinburgh*. University Press.

Fang, N., & Guo, Y. (2013). A web-based interactive intelligent tutoring system for undergraduate engineering dynamics. In 2013 IEEE Frontiers in Education Conference (FIE) (pp. 159–160). IEEE. doi:10.1109/FIE.2013.6684806

Fang, N. & Guo, Y. (2016). Interactive computer simulation and animation for improving student learning of particle kinetics. Journal of Comp. Assisted Learning, 32(5). 10.1111/jcal.12145

Felder, R. M. (1996). Matters of style. *ASEE Prism*, *6*(4), 18–32.

Felder, R. M., & Brent, R. (2005). Understanding student differences. *Journal of Engineering Education*, *94*(1), 57–72. doi:10.1002/j.2168-9830.2005.tb00829.x

Felder, R. M., & Brent, R. (2005a). Understanding Student Differences. *Journal of Engineering Education*, *94*(1), 57–72. doi:10.1002/j.2168-9830.2005.tb00829.x

Felder, R. M., & Silverman, L. K. (1988). Learning styles and teaching styles in engineering education. *Journal of Engineering Education*, *78*(7), 674–681.

Felder, R. M., & Spurlin, J. (2005). Applications, reliability and validity of the index of learning styles. *International Journal of Engineering Education*, *21*(1), 103–112.

Ferdig, R. E. (2006). Assessing technologies for teaching and learning: Understanding the importance of technological pedagogical content knowledge. *British Journal of Educational Technology*, *37*(5), 749–760. doi:10.1111/j.1467-8535.2006.00559.x

Fleming, N. D. (1995). I'm different; not dumb: Modes of presentation (VARK) in the tertiary classroom. In A. Zelmer (Ed.), *Proceedings of the 1995 Annual Conference of the Higher Education and Research Development Society of Australasia (HERDSA)* (Vol. 18, pp. 308–313). Queensland, Australia: Research and Development in Higher Education. Retrieved from http://vark-learn.com/wp-content/uploads/2014/08/different_not_dumb.pdf

Fleming, N. D. (2005). *Teaching and learning styles: VARK strategies* (2nd ed.). Christchurch, New Zealand: The Digital Print and Copy Center.

Fosnot, C. (1996). Constructivism: A psychological theory of learning. In C. Fosnot (Ed.), *Constructivism: Theory, perspectives, and practice* (pp. 8–33). New York, NY: Teachers College Press.

Froyd, J. E., Wankat, P. C., & Smith, K. A. (2012). Five major shifts in 100 years of engineering education. *Proceedings of the IEEE*, *100*(Special Centennial Issue), 1344–1360. doi:10.1109/JPROC.2012.2190167

Gagne, R. M. (1985). *The Conditions of Learning and Theory of Instruction* (4th ed.). Wadsworth Pub Co.

Gao, Z., Liu, S., Ji, M., & Liang, L. (2011). Virtual hydraulic experiments in courseware: 2D virtual circuits and 3D virtual equipments. *Computer Applications in Engineering Education*, *19*(2), 315–326. doi:10.1002/cae.20313

Graf, S. (2007). *Adaptivity in learning management systems focusing on learning styles* (Doctoral dissertation). Vienna University of Technology.

Graf, S., Viola, S. R., Leo, T., & Kinshuk. (2007). In-depth analysis of the Felder-Silverman learning style dimensions. *Journal of Research on Technology in Education*, *40*(1), 79–93. doi:10.1080/15391523.2007.10782498

Gregorc, A. F. (1985). *Inside Styles: Beyond the Basics*. Columbia, CT: Gabriel Systems.

Gupta, A. (2002). Content development for e-learning in engineering education. *Interactive Educational Multimedia*, *4*, 12–23.

Hannafin, M. J., & Land, S. M. (1997). The foundations and assumptions of technology enhanced student-centered learning environments. *Instructional Science*, *25*(3), 167–202. doi:10.1023/A:1002997414652

Haque, E., Nayna, T. F. A., & Ahmed, F. (2016). A pragmatic approach for quality enhancement in classroom teaching of engineering education in developing countries: A case study on power electronics. In 2016 Future Technologies Conference (FTC) (pp. 349–354). IEEE. doi:10.1109/FTC.2016.7821633

Hawk, T. S., & Shah, A. J. (2007). Using Learning Style Instruments to Enhance Student Learning. *Decision Sciences Journal of Innovative Education*, *5*(1), 1–19. doi:10.1111/j.1540-4609.2007.00125.x

Hmelo, C. E., Lunken, E. Y., Gramoll, K., & Yusuf, I. (1995). Multimedia courseware for teaching dynamic concepts: Assessment of student learning. In *Proceedings of the Frontiers in Education Conference* (pp. 19-23). Piscataway, NJ: IEEE Press. 10.1109/FIE.1995.483047

Hogstad, N. M., Isabwe, G. M. N., & Vos, P. (2016). *Engineering students' use of visualizations to communicate about representations and applications in a technological environment*. In *First conference of International Network for Didactic Research*, Montpellier, France.

Holvikivi, J. (2007). Learning styles in engineering education: The Quest to improve didactic practices. *European Journal of Engineering Education, 32*(4), 401–408. doi:10.1080/03043790701332909

Hone, J. (2012). *What is Mechanical Engineering?* Retrieved April 14, 2017, from http://me.columbia.edu/what-mechanical-engineering

Honey, P., & Mumford, A. (1992). *The Manual of learning styles (3rd ed.).* Maidenhead, UK: Peter Honey Publications.

Hsieh, S. H., Chen, C. S., & Lu, C. C. (2005). Interactive tutoring components for engineering dynamics. In S.-H. Ou (Ed.), *Proceedings of the iCEER-2005 iNEER Conference for Engineering Education and Research.* Tainan, Taiwan: International Network for Engineering Education and Research (iNEER).

Huczynski, A., & Johnston, S. P. (2005). Engineering students' use of computer assisted learning (CAL). *European Journal of Engineering Education, 30*(2), 287–298. doi:10.1080/03043790500087530

Janet, F. J. (2002). Sequence Learning with Intelligent 3-D Practice Environments. *MITRE Technical Papers*, Paper 2.

Janson, J. L. (1992). Computer-based training helps firms trim budgets. *PC World*, 56–61.

John, A. R., & Abdollah, A. (1998). Developing interactive educational engineering software for the World Wide Web with java. *Computers in Education Journal, 30*(3/4), 183–194.

Jonassen, D. H. (1991). Objectivism vs constructivism: Do we need a new philosophical paradigm. *Educational Technology Research and Development, 39*(3), 5–14. doi:10.1007/BF02296434

Kabouridis, G. 2010. An innovative approach to face the first-year mechanical engineering students' conceptual difficulties in engineering design. *Proceedings of 1st WIETE Annual Conference on Engineering and Technology Education*, 110-113. Retrieved from www.wiete.com.au/conferences/1st_wiete/20-05-Kabouridis.pdf

Kay, R. H., & Knaack, L. (2008). A multi-component model for assessing learning objects: The learning object evaluation metric (LOEM). *Australasian Journal of Educational Technology, 24*(5), 574–591. doi:10.14742/ajet.1192

Keefe, J. W. (1991). *Learning style: Cognitive and thinking skills.* Reston, VA: National Association of Secondary School Principals.

Kek, M. Y. C. A., & Huijser, H. (2011). The power of problem-based learning in developing critical thinking skills: Preparing students for tomorrow's digital futures in today's classrooms. *Higher Education Research & Development, 30*(3), 329–341. doi:10.1080/07294360.2010.501074

Kelting-Gibson, L. (2013). Analysis of 100 Years of Curriculum Designs. *International Journal of Instruction, 6*(1), 39–58.

Kirkpatrick, D. L., & Craig, R. (1976). *Evaluation in the ASTD training and development handbook: A guide to human resource development (4th ed.).* New York: McGraw-Hill.

Kirkwood, A. (2009). E-learning: You don't always get what you hope for. *Technology, Pedagogy and Education, 18*(2), 107–121. doi:10.1080/14759390902992576

Kirschner, P. A., & Erkens, G. (2006). Cognitive tools and mindtools for collaborative learning. *Journal of Educational Computing Research, 35*(2), 199–209. doi:10.2190/R783-230M-0052-G843

Ko, S., & Rossen, S. (2005). *Teaching online: A practical guide* (2nd ed.). New York: Houghton Mifflin.

Kojmane, J., & Aboutajeddine, A. (2016). Enjoyeering Junior: A hands-on activity to enhance technological learning in an engineering dynamics course. In *2016 International Conference on Information Technology for Organizations Development (IT4OD)* (pp. 1–6). IEEE. 10.1109/IT4OD.2016.7479267

Kolb, D. A. (1983). *Experiential Learning: Experience as the Source of Learning and Development.* New York, NY: Financial Times/ Prentice Hall.

Kolb, D. A., Boyatzis, R. E., & Mainemelis, C. (2001). Experiential learning theory: Previous research and new directions. In R. J. Sternberg & L. Zhang (Eds.), *Perspectives on thinking, learning and cognitive styles* (pp. 227–247). London, UK: Lawrence Erlbaum Associates.

Kulik, C. C., & Kulik, J. A. (1991). Effectiveness of computer-based instruction: An updated analysis. *Computers in Human Behavior, 7*(1-2), 75–94. doi:10.1016/0747-5632(91)90030-5

Kuri, N. P., & Truzzi, O. M. S. (2002). Learning styles of freshmen engineering students. In *Proc. 2002 International Conference on Eng. Education*. Arlington, VA: International Network for Engineering Education and Research (iNEER).

Lai, K.-W. (1993). Teachers as facilitators in a computer-supported learning environment. *Journal of Information Technology for Teacher Education*, *2*(2), 127–137. doi:10.1080/0962029930020202

Larkin-Hein, T., & Budny, D. D. (2001). Research on learning style: Applications in the physics and engineering classrooms. *IEEE Transactions on Education*, *44*(3), 276–281. doi:10.1109/13.941000

Lau, H. Y. K., Mak, K. L., & Helen, M. (2006). IMELS: An e-learning platform for industrial engineering. *Journal of Computer Applications in Engineering Education*, *14*(1), 53–63. doi:10.1002/cae.20067

Lee, C. K., & Manjit Sidhu, S. (2013). Engineering students learning styles preferences using Honey and Mumford learning styles questionnaire: A Case study in Malaysia. *International Journal of Information Technology & Computer Science*, *9*(1), 107–114.

Liang, H. N., & Sedig, K. (2010). Can interactive visualization tools engage and support preuniversity students in exploring non-trival mathematical concepts? *Computers & Education*, *54*(4), 972–991. doi:10.1016/j.compedu.2009.10.001

Liarokapis, F., Mourkoussis, N., White, M., Darcy, J., Sifniotis, M., Petridis, P., & ... . (2007). Web3D and augmented reality to support engineering education. *Engineering and Technology Education*, *3*(1), 1–4.

Lindsay, G. (2007). Educational psychology and the effectiveness of inclusive education/mainstreaming. *British Journal of Educ*.

Livesay, G. A., Dee, K. C., Nauman, E. A., & Hites, L. S. (2002). Engineering student learning styles: A Statistical analysis using Felder's Index of Learning Styles. In P*roceeding of the 2002 Annual Conference of the American Society for Engineering Education*. Montreal, Quebec: ASEE.

Lopez, W. M. (2002). *ILS -Inventario de aprendizagem de Felder-Solomon: Investigaçao du sua validade em estudentes Universitarios de Belo Horizante* [ILS-Inventory of Felder-Solomon Learning Styles: Research validity of students in University of Belo Horizonte] (Unpublished masters dissertation). Universidade Federal de santa Catarina, Brazil.

Lynn, E. K., Lenly, J., Weathers, W. M., Ronal, K., Kevin, A., & Mike, S. (1997). A multimedia based laboratory course for environmental engineering fundamentals and process design. *Proceedings of the 27th Annual Conference of Teaching and Learning in an Era of Change*, 748. doi:10.1109/FIE.1997.635931

Mackenzie, J. G., Earl, W. B., Allen, R. M., & Gilmour, I. A. (2001). Amoco computer simulation in chemical engineering education. *Journal of Engineering Education*, *90*(3), 331–343. doi:10.1002/j.2168-9830.2001.tb00612.x

Manjit Sidhu, S. (2006). *Development and applications of interactive multimedia TAPS (Technology Assisted Problem Solving) packages for engineering* (PhD Thesis). University Malaya, Malaysia.

Manjit Sidhu, S. (2009). *Technology-Assisted Problem Solving for Engineering Education*. Hershey, PA: IGI Global. doi:10.4018/978-1-60566-764-5

Manjit Sidhu, S., Ramesh, S., & Selvanathan, N. (2004). The multimedia approach in visualizing engineering concept. *International Journal of Computing & Information Science*, *2*(3), 129–136.

Marsico, S. A. (2001). Experience with the introduction of multimedia into mechanical engineering technology, mechanics of materials laboratory. *Proceedings of the 2001 American Society for Engineering Education Annual Conference & Exposition*, 1–10. Retrieved from http://search.asee.org/search/fetch?url=file%3A%2F%2Flocalhost%2FE%3A%2Fsearch%2Fconference%2F25%2FAC%25202001Paper383.PDF&index=conference_papers&space=129746797203605791716676178&type=application%2Fpdf&charset=

Martin, S., Diaz, G., Sancristobal, E., Gil, R., Castro, M., & Peire, J. (2011). New technology trends in education: Seven years of forecasts and convergence. *Computers & Education*, *57*(3), 1893–1906. doi:10.1016/j.compedu.2011.04.003

Mautone, P. D., & Mayer, R. E. (2001). Signaling as a cognitive guide in multimedia learning. *Journal of Educational Psychology*, *93*(2), 377–389. doi:10.1037/0022-0663.93.2.377

Mayer, R. E., & Moreno, R. (2003). Nine ways to reduce cognitive load in multimedia learning. *Educational Psychologist*, *38*(1), 43–52. doi:10.1207/S15326985EP3801_6

McCaulley, M. H. (2000). Myers-Briggs type indicator: A Bridge between counseling and consulting. *Consulting Psychology Journal: Practice and Research*, *52*(2), 117–132. doi:10.1037/1061-4087.52.2.117

McMahon, G. (2009). Critical thinking and ICT integration in a Western Australian secondary school. *Journal of Educational Technology & Society*, *12*(4), 269–281.

Mehendale, S., & Irwin, J. L. (2015). An Interactive Dynamics Learning Course (IDLC). *122nd ASEE Annual Conference & Exposition.*

Miri, B., David, B., & Uri, Z. (2007). Purposely teaching for the promotion of higher-order thinking skills: A case of critical thinking. *Research in Science Education*, *37*(4), 353–369. doi:10.100711165-006-9029-2

Miskioglu, E. E., & Wood, D. W. (2013). Learning in style: Correlation of learning styles with problem comprehension and perceptions in an introductory chemical engineering course. In *Proceedings of 2013 IEEE Frontiers in Education Conference (FIE)* (pp. 979–981). IEEE. 10.1109/FIE.2013.6684973

Mistree, F., Panchal, J. H., Schaefer, D., Allen, J. K., Haroon, S., & Siddique, Z. (2014). Personalized Engineering Education for the Twenty-First Century. In M. Gosper & D. Ifenthaler (Eds.), Curriculum Models for the 21st Century (1st ed.; pp. 91–111). Academic Press. doi:10.1007/978-1-4614-7366-4_6

Montgomery, S. M. (1995). Addressing diverse learning styles through the use of multimedia. In *Proceedings Frontiers in Education 1995 25th Annual Conference. Engineering Education for the 21st Century* (Vol. 1, pp. 3a2.13–3a2.21). Academic Press. 10.1109/FIE.1995.483093

Morsi, R., & Mull, S. (2015). Digital Lockdown: A 3D adventure game for engineering education. In 2015 IEEE Frontiers in Education Conference (FIE) (pp. 1–4). IEEE. doi:10.1109/FIE.2015.7344072

Nasr, K. J. (2014). *Towards A Converged and Global Set Of Competencies For Graduates of Engineering Programs In A Globalization-Governed World* (18). Academic Press.

National Academies of Science, Engineering, and Medicine. (2007). *Rising above the gathering storm: Energizing and employing America for a brighter economic future*. Washington, DC: National Academies Press.

National Science Board. (2007). *Moving forward to improve engineering education* (NSB-07-122). Retrieved from www.nsf.gov/pubs/2007/nsb07122/nsb07122.pdf

Noguera, J. S., & Wageman, J. (2011). Spanish EFL undergraduate students' perceptions of learning styles. *Nordic Journal of English Studies*, *10*(1), 77–101. doi:10.35360/njes.243

Norman, D. A. (1993). *Things that make us smart: Defending human attributes in the age of the machine*. New York, NY: Addison-Wesley.

Noronha, M., Bittencourt, T. N., Proença, S. P., & Guello, G. A. (2000). Multimedia-based environment in structural engineering education. In C. da R. Brito (Ed.), *The II International Conference on Engineering and Computer Education – Cooperative Network for Engineering and Computer Education Development* (pp. 1–5). Sao Paulo, Brazil: Science and Education Research Council. Retrieved from http://www.lmc.ep.usp.br/pesquisas/TecEdu/artigos/icece00.pdf

Norrie, S. E. (1996). Evaluation of computer based laboratory simulation. *Computers in Education Journal*, *26*(1-3), 123–130. doi:10.1016/0360-1315(95)00057-7

Ogot, M., & Okudan, G. L. E. (2007). Systematic creativity methods in engineering education: A Learning styles perspective. *International Journal of Engineering Education*, *22*(3), 566–576.

Palou, E. (2006). Learning Styles of Mexican Food Science and Engineering Students. *Journal of Food Science Education*, *5*(3), 51–57. doi:10.1111/j.1541-4329.2006.00006.x

Park, I., & Hannafin, M. J. (1993). Empirically based guidelines for the design of interactive multimedia. *Educational Technology Research and Development*, *41*(3), 63–85. doi:10.1007/BF02297358

Paterson, K. G. (1999). Student perceptions of internet-based learning tools in environmental engineering education. *Journal of Engineering Education*, *88*(3), 295–304. doi:10.1002/j.2168-9830.1999.tb00450.x

Patterson, D. A. (2011). Impact of a multimedia laboratory manual: Investigating the influence of student learning styles on laboratory preparation and performance over one semester. *Education for Chemical Engineers*, *6*(1), e10–e30. doi:10.1016/j.ece.2010.10.001

Pedrosa, C. M., Barbero, B. R., & Miguel, A. R. (2014). Spatial visualization learning in engineering: Traditional methods vs. a Web based tool. *Journal of Educational Technology & Society*, *17*(2), 142–157.

Pedrosa de Jesus, H., Almeida, P. C., & Watts, M. (2004). Questioning styles and students' learning: Four case studies. *Educational Psychology*, *24*(4), 531–548. doi:10.1080/0144341042000228889

Pendleton, D. (1984). *The Consultation: An Approach to Learning and Teaching (Oxford General Practice)*. Oxford, UK: Oxford University Press.

Pittenger, D. J. (2005). Cautionary comments regarding the Myers-Briggs type indicator. *Consulting Psychology Journal: Practice and Research*, *57*(3), 210–221. doi:10.1037/1065-9293.57.3.210

Prados, J. W. (1998). Action agenda for systematic engineering education reform: Next steps. In *Proceedings of the Conference on Realizing the New Paradigm for Engineering Education, Engineering Foundation Conference* (pp. 1–13). IEEE.

Prince, M. (2004). Does Active Learning Work? A Review of the Research. *Journal of Engineering Education*, *93*(3), 223–231. doi:10.1002/j.2168-9830.2004.tb00809.x

Rajala, S. A. (2012). Beyond 2020: Preparing engineers for the future. In *Proceedings of the IEEE 100 (Centennial-Issue)* (pp. 1376–1383). IEEE. 10.1109/JPROC.2012.2190169

Ranga, K., & Gramoll, K. (1999). Design education over the internet using VRML. In *Proceedings of the Society Engineering Education (ASEE) 1999 Conference. Session 2625* (pp. 1–10). ASEE. Retrieved from http://eml.ou.edu/paper/paper/1999

Ré, M. A., & Giubergia, M. F. (2013). Virtual Laboratory for a first experience in dynamics. In *ICBL2013 – International Conference on Interactive Computer aided Blended Learning* (pp. 239–244). Academic Press.

Reiser, R. (2001). A history of instructional design and technology: Part 1: A history of instructional media. *Educational Technology Research and Development*, *49*(1), 53–64. doi:10.1007/BF02504506

Renes, S. L., & Strange, A. T. (2011). Using technology to enhance higher education. *Innovative Higher Education*, *36*(3), 203–213. doi:10.100710755-010-9167-3

Renes, S. L., & Strange, A. T. (2011). Using technology to enhance higher education. *Innovative Higher Education, 36*(3), 203–213.

Riding, R. J., & Rayner, S. G. (1998). *Cognitive styles and learning strategies: Understanding style differences in learning and behavior*. London, UK: David Fulton Publishers.

Rieber, L., & Kini, A. S. (1991). Theoretical foundations of instructional applications of computer-generated animated visuals. *Journal of Computer-Based Instruction, 18*(3), 83–88.

Ringstaff, C., & Kelley, L. (2002). *The learning return on our educational technology investment*. San Francisco, CA: WestEd.

Rosati, P. (1999). Specific differences and similarities in the learning preferences of engineering students. In *Proceedings of Frontiers in Education (FIE)'99, 29th Annual Frontiers in Education Conference* (Vol. 2, pp. 12C1/17–12C1/22). IEEE. 10.1109/FIE.1999.841625

Rosati, P. A. (1996). Comparisons of learning preferences in an engineering program. In *Technology-Based Re-Engineering Engineering Education Proceedings of Frontiers in Education (FIE)'96 26th Annual Conference* (Vol. 3, pp. 1441–1444). Academic Press. 10.1109/FIE.1996.568536

Rosen, M. A. (2007). Future trends in engineering education. In Innovations 2007: World Innovations in Engineering Education and Research (pp. 1–11). Arlington, VA: International Network for Engineering Education and Research (iNEER).

Ross, S. M., Morrison, G. R., & Lowther, D. L. (2010). Educational Technology Research Past and Present: Balancing Rigor and Relevance to Impact School Learning. *Contemporary Educational Technology, 1*(1), 17–35.

Sadler-Smith, E. (1997). "Learning Style": Frameworks and instruments. *Educational Psychology, 17*(1-2), 51–63. doi:10.1080/0144341970170103

Sakamotao, T., Kimurab, K., Shimadac, M., Ozawad, S., & Okamoto, T. (1979). Evaluation methods for improving CAL materials. *Computers & Education, 3*(4), 281–291. doi:10.1016/0360-1315(79)90005-8

Salam, S. (2011). *Investigation of students' learning dynamics in an augmented reality environment to support individualisation, collaboration and creativity*. Academic Press.

Salvatore, A. M. (2001). Experience with the Introduction of Multimedia into Mechanical Engineering Technology, Mechanics of Materials Laboratory. *Proceedings of the 2001 American Society for Engineering Education Society*, 1 – 10.

Schar, S., Schluep, S., Schierz, C., & Krueger, H. (2000). Interaction for computer-aided learning. *Interactive Multimedia Electronic Journal of Computer-Enhanced Learning, 2*(1). Retrieved from http://imej.wfu.edu/articles/2000/1/03/

Scott, N. W. (1996). *A study of the introduction of educational technology into a course in engineering dynamics: Classroom environment and learning outcomes* (Unpublished doctoral dissertation). University of Western Australia, Perth, Australia.

Seery, N., Gaughran, W. F., & Waldmann, T. (2003). Multi-modal learning in engineering education. In *Proceedings of 2003 Annual ASEE Conference*. Nashville, TN: ASEE.

Self, B. P., & Widmann, J. (2009). Work in progress - learning styles and performance on the dynamics concept inventory. In 2009 39th IEEE Frontiers in Education Conference (pp. 1–2). IEEE. 10.1109/FIE.2009.5350822

Shang, H. H. Chuin, S. C., & Cho, C. L. (2005). Interactive Tutoring Components for Engineering Dynamics. *Exploring Innovation in Education and Research.*

Shapiro, A. M. (1998). Promoting Active Learning: The Role of System Structure in Learning From Hypertext. *Huamn-Computer Interaction, 13*(1), 1–35.

Smeets, E. (2005). Does ICT contribute to powerful learning environments in primary education? *Computers & Education, 44*(3), 343–355. doi:10.1016/j.compedu.2004.04.003

Smith, N. G., Bridge, J., & Clarke, E. (2002). An evaluation of students' performance based on their preferred learning styles. In Z. J. Pudlowski (Ed.), *Proc. UICEE Annual Conference on Engineering Education*. UNESCO International Centre for Engineering Education.

Squires, D. (1999). Usability and educational software design: Special issue of interacting with computers. *Interacting with Computers, 11*(5), 463–466. doi:10.1016/S0953-5438(98)00062-9

Stanley, R., & DiGiuseppe, G. (2009). An Efficient Way to Increase the Engineering Student's Fundamental Understanding of Thermodynamics by Utilizing 227 Interactive Web Based Animation Software. *Proceedings of the 2009 ASEE Annual Conference & Exposition.*

Stemler, L. K. (1997). Educational characteristics of multimedia: A literature review. *Journal of Educational Multimedia and Hypermedia*, *6*(3-4), 339–359.

Sternberg, R. J., & Grigorenko, E. L. (1997). Are cognitive styles still in style? *The American Psychologist*, *52*(7), 700–712. doi:10.1037/0003-066X.52.7.700

Stice, J. E. (1987). Using Kolb's Learning Cycle to Improve Student Learning. *Journal of Engineering Education*, *77*(5), 291–296.

Stone, D., Jarrett, C., Woodroffe, M., & Minocha, S. (2005). *User interface design and evaluation*. San Francisco, CA: The Open University.

Tamim, R. M., Lowerison, G., Schmid, R. F., Bernard, R. M., & Abrami, P. C. (2011). A multiyear investigation of the relationship between pedagogy, computer use and course effectiveness in postsecondary education. *Journal of Computing in Higher Education*, *23*(1), 1–14. doi:10.100712528-010-9041-4

Tembe, B. L., & Kamble, S. K. (2016). Implementation of Active Learning Methods in Mechanical Engineering Education to Enhance Students' Performance. In 2016 IEEE 40th Annual Computer Software and Applications Conference (COMPSAC) (pp. 258–263). IEEE. 10.1109/COMPSAC.2016.12

Tendy, S. M., & Geiser, W. F. (1997). The Search for style: It all depends on where you look. *National FORUM of Teacher Education Journal*, *9*(1), 3–15.

Teoh, B. S. P., & Neo, T. K. (2006). Innovative teaching: Using multimedia to engage students in interactive learning in higher education. In *Proceedings of the 7th International Conference on Information Technology Based Higher Education and Training, ITHET* (pp. 329–337). Academic Press. 10.1109/ITHET.2006.339782

The American Heritage Dictionaries (2011). The American Heritage Dictionary of the English Language (5th ed.). Houghton Mifflin Harcourt Trade.

Tinio, V. (2003). *ICT in education*. New York, NY: UNDP-APDIP. Retrieved from http://www.saigontre.com/FDFiles/ICT_in_Education

Tsai, C.-C., Chai, C. S., Wong, B. K. S., Hong, H.-Y., & Tan, S. C. (2013). Positioning Design Epistemology and its Applications in Education Technology. *Journal of Educational Technology & Society*, *16*(2), 81–90.

Tway, L. (1995). *Multimedia in action*. AP Professional.

Urhahne, D., Schanze, S., Bell, T., Mansfield, A., & Holmes, J. (2010). Role of the teacher in computer-supported collaborative inquiry learning. *International Journal of Science Education*, *32*(2), 221–243. doi:10.1080/09500690802516967

Vita, G. V. (2001). Learning styles, culture and inclusive instruction in the multicultural classroom: A Business and management perspective. *Innovations in Education and Teaching International*, *38*(2), 165–174. doi:10.1080/14703290110035437

Wang, M., Shen, R., Novak, D., & Pan, X. (2009, July). students' learning behaviours and performance: Report from a large blended classroom. *British Journal of Educational Technology*, *40*(4), 673–695. doi:10.1111/j.1467-8535.2008.00846.x

Welty, G. (2007). The "Design" Phase of the ADDIE Model. *Journal of GXP Compliance*, *11*(4), 40–48.

Wilson, L. (2002). Using Multimedia in Teaching Dynamics. *Proceedings of the 2002 Conference for Industry and Education Collaboration*.

Wince-Smith, D. (2005). Innovate America: Thriving in a world of challenge and change. In *Proceedings of the Global Innovation Ecosystem 2007 Symposium* (pp. 1-12). Academic Press. Retrieved from http://crds.jst.go.jp/GIES/archive/GIES2007/en/symposium/

Zheng, R., Snelbecker, G., Miller, S., & Cohen, I. (2006). Use of multimedia for problem-solving tasks. *Journal of Technology, Instruction. Cognition and Learning*, *3*(1-2), 135–143.

Zheng, R., Yang, W., Garcia, D., & McCadden, E. (2008). Effects of multimedia and schema induced analogical reasoning on science learning. *Journal of Computer Assisted Learning*, *24*(6), 474–482. doi:10.1111/j.1365-2729.2008.00282.x

Zhu, C. (2010). Teacher roles and adoption of educational technology in the Chinese context. *Journal for Educational Research Online*, *2*(2), 72–86.

Zohar, A. (2004). Elements of teachers' pedagogical knowledge regarding instruction of higher order thinking. *Journal of Science Teacher Education, 15*(4), 293–312. doi:10.1023/B:JSTE.0000048332.39591.e3

Zywno, M. S. (2002). Instructional technology, learning styles and academic achievement. In *Proceedings of the 2002 American Society for Engineering Education Annual Conference & Exposition*. Montreal, Canada: ASEE.

Zywno, M. S. (2003). *The Effect of individual differences on learning outcomes through hypermedia-based instruction* (Doctoral dissertation). Glasgow Caledonian University, Glasgow, UK.

Zywno, M. S., & Waalen, J. K. (2001). The Effect of hypermedia instruction on achievement and attitudes of students with different learning styles. In *Proceedings of the 2001 American Society for Engineering Education Annual Conference & Exposition*. Albuquerque, NM: ASEE.

KEY TERMS AND DEFINITIONS

CAL: Computer-aided learning is an integrative technology, which describes an educational environment where a computer program is used to assist the user in learning a particular subject. It refers to an overall integrated approach of instructional methods.

Felder-Silverman Model: A multifaceted approach to understanding learning style. This model includes: active and reflective learners, sensing and intuitive learners, visual and verbal learners, and sequential and global learners.

Honey and Mumford Learning Styles: Learning style inventory which is based on the Kolb learning styles, but with a few differences.

Kolb's Learning Style Model: Learning style inventory that works on two levels: a four-stage cycle of learning and four separate learning styles which is concerned with the learner's internal cognitive processes.

Learning Styles: Learning styles group common ways that people learn. Everyone has a mix of learning styles. Some people may find that they have a dominant style of learning, with far less use of the other styles. Others may find that they use different styles in different circumstances.

Multimedia: Multimedia is the field concerned with the computer-controlled integration of text, graphics, drawings, still and moving images (Video), animation, audio, and any other media where every type of information can be represented, stored, transmitted, and processed digitally.

Myers-Briggs Type Indicator (MBTI): Is an introspective self-report questionnaire with the purpose of indicating differing psychological preferences in how people perceive the world around them and make decisions.

Chapter 3
Interface Design

ABSTRACT

User interface (UI) design is the process of making interfaces in software or computerized devices with a focus on looks or style. Designers aim to create designs users will find easy to use and pleasurable. IU design typically refers to graphical user interfaces but also includes others, such as voice-controlled ones. In this chapter, the user interface design and the grounded learning theories are discussed. Next, the interaction styles and the types of interactions are discussed. The usability benchmark and the usability evaluation instruments are also discussed in this chapter.

USER INTERFACE DESIGN

User interface is defined as the part of computer system with which a user interacts in order to undertake his or her tasks and achieve his or her goal (Stone et al., 2005). It is the bridge between the world of computer system and the world of the user. It is widely acknowledged by researchers (Norman, 2013; Shneiderman et al., 2017) that good interfaces ought to satisfy the principle of visibility: (a) users should be able to 'see' the actions that are open to them at every choice point, (b) they should receive immediate feedback about the actions they have just taken – since few things upset computer users more than not knowing what a computer is doing when it seems to be churning

DOI: 10.4018/978-1-7998-0465-9.ch003

unexpectedly, and (c) they should get timely and insightful information about the consequences of their actions.

Unfortunately, many system developers have over emphasized on the technical functionalities of the user interface design thus neglected the usability aspects of the interface design. These efforts violate the initial purpose of the interface design that user is essential in this field (Atoum & Bong, 2015; Maguire, 2013; Manresa-Yee et al., 2010; Norman, 1999). As stated by Luostarinen et al. (2010), the user interface design is often heavily based on the technical properties of the devices thus cause the usability related issues to be forgotten or totally dismissed in the design process. This is in line with the argument by Maguire (2013) that many systems are hard to use which leads to poor user experience by causing people to abandon the system or fail to use the system effectively. Therefore, the goals for a good human-computer interaction are to decrease the errors, increase satisfaction for the user, and better performance of machine-assisted tasks (Manresa-Yee et al., 2010).

Designing a user interface with good usability is a challenging task. Fortunately, many interface design practitioners did propose the guidelines for good interface design. Shneiderman's Eight Golden Rules of Interface Design suggests the following guidelines for the practitioner as follow (Shneiderman et al., 2017):

1. Strive for consistency
2. Enable frequent users to use shortcuts
3. Offer information feedback
4. Design dialog to yield closure
5. Offer simple error handling
6. Permit easy reversal of actions
7. Support internal locus of control
8. Reduce short-term memory load

Jacob Nielsen's ten usability heuristics also serve as the basic reference for all user interface design (Nielsen, 1995) listed in Table 1. Nielsen stressed in his article that these ten principles are called "heuristic" because they should be treated as broad rules of thumb not specific usability guidelines. Although the Nielsen and Shneiderman's interface design guidelines can act as the generic reference for all the user interface system design, however, Gong

Table 1. Ten usability heuristics for interface design (Adopted from Nielsen, 1995)

10 Usability Heuristics for User Interface Design	
Visibility of system status	Recognition rather than recall
Match between system and the real world	Flexibility and efficiency of use
User control and freedom	Aesthetic and minimalist design
Consistency and standards	Help users recognize, diagnose, and recover from errors
Error prevention	Help and documentation

(2009) raised the concern more specifically that additional efforts must be included to identify different types of users and their needs and skills, and to analyze different tasks and the purpose of the system while designing a user interface. This is consistent with the statement as argued by Obeidat and Salim (2010) that different users have different characteristics, such as background, education, personality, cognitive skills and preferences, thus designing a good interface would be a great challenge.

The next section will further discuss the interface design issues specifically related to the computer aided learning environments

USER INTERFACE DESIGN FOR COMPUTER AIDED LEARNING ENVIRONMENT

In the knowledge age, the requirements for effective learning have changed as compare to previous era. The five key requirements for effective learning in the knowledge age are summarized in Table 2.

Educational technologies plays a major role in the learning science. As raised by Oviatt (2013), the current research focus is on how technology can be used to support the requirements as stated in Table 2. Similarly, the challenge on how the interface design may use to facilitate the users learning process in the computer aided learning environment is yet to empirically explore. As noted by Peters (2014):

"Research has now proven that multimedia and interface design affects how users learn. The myth of visual design as an optional extra is in desperate need of busting. The hard fact is that how you create graphics, sequence interaction, display information, use animation, and design for social presence and emotion will impact how users learn. This is interface design. And this is where a new sub discipline is ready to step in and change the game."

Table 2. Five key requirements for effective learning (Adapted from Sawyer, 2008)

	Key Requirements	Descriptions
1	Support deep conceptual understanding	Expert knowledge includes facts and procedures, but acquiring these isn't enough. Facts and procedures are useful only when a person knows when and how to apply them, and how to adapt them to new contexts.
2	Focus on learning, not just teaching	Students can gain deep conceptual understanding only by actively participating in their own learning process. The learning sciences focus on students' learning processes as well as teachers' instructional techniques.
3	Create learning environments	The role of schools is to support students in becoming competent adult experts. This includes learning facts and procedures, but also gaining the deeper conceptual understanding necessary for real-world problem solving.
4	Build on learners' prior knowledge	Student learn best from experiences that build on their existing knowledge, which includes working with both accurate and flawed preconceptions.
5	Support reflection	Learners benefit from opportunities to express their developing knowledge and to analyse their current state of understanding, whether through discussion or the creation of artefacts like papers, reports or media.

In response to the research trend that increasingly emphasized on the important role of interface design for learning, Oviatt (2013) stressed on the following key points for further exploration:

1. New educational interfaces need to be designed as cognitive tools for stimulating flexible thinking and adaptive learning rather than formatting documents and transmitting information.
2. New educational interfaces are needed that increase students' engagement in complex, multi-part, extended problems rather than supporting the simple and repetitive tasks.

Thus, the researchers did highlight the difference between the user-centered design and learner-centered design. As mentioned by Quintana et al (2006), developing new understanding is the main goal for learners. Thus, the indicator for a successful learner-centered software is that the learner's understanding will grow and change significantly while using the software. While user-centered design is often about supporting task completion by users, learner-centered design is about transforming the user (Peters, 2014).

The design and development of educational interfaces should be grounded on learning theories as the foundation. The traditional learning theories (behaviorism, cognitivism, constructivism) and the digital age learning theory (connectivism) will be further discussed in the next section.

LEARNING THEORY FOR EDUCATIONAL INTERFACES

Learning theories are conceptual frameworks describing how knowledge is absorbed, processed and retained during learning (Simandan, 2013). As stated by Ertmer & Newby (2013), learning is a complex process that has generated numerous interpretation and theories of how it is effectively accomplished. The common learning theories that are widely applicable in the context of educational technology are behaviourism, cognitivism, constructivism and connectivism. Different learning theories are grounded based on different philosophy. According to the famous quotation by Hilgard and Bower (1966), they described that:

"While it is extremely difficult to formulate a satisfactory definition of learning so as to include all the activities and processes which we wish to include and eliminate all those which we wish to exclude the difficulty does not prove to be embarrassing because it is not a source of controversy as between theories. The controversy is over fact and interpretation, not over definition."

Peters (2014) further add on that different perspective on learning are valuable for different reasons and within different contexts, and no single theory is ideal for every situation. This is due to the reason that learning is a complex process that normally is not performed in a linear fashion. Furthermore, each learning theory has its strengths and weaknesses, thus the selection of the theory that suits for certain learning context depends on multiple aspects of learning such as learning goals, learners and situation (Arshavskiy, 2013). The task of translating the learning theory into practical applications for the interface design is a challenging task that needs further exploration. The common learning theories would be reviewed and discussed in the next section.

Behaviorism

Behaviourism is a philosophy, theory and pedagogy that are underpinned by the principles of stimulus-response, without the focus on any attempt to consider internal mental processes (Peters, 2014; Winn, 1990; Woollard, 2011). Behaviorism is based on observable and measureable changes in behavior (Arshavskiy, 2013). Behaviorists claimed that only observable behavior was worth studying (to measure, predict, and manipulate patterns of behavior) empirically (Peters, 2014). The learner is characterized as being reactive to conditions in the environment as opposed to taking an active role in discovering the environment (Ertmer & Newby, 2013). The behaviorist strategies have generally been proven reliable and effective in facilitating learning that involves recalling facts, defining and illustrating concepts, applying explanations, and automatically performing a specified procedure (Ertmer & Newby, 2013). However, it is generally agreed that behavioral principles cannot adequately explain the acquisition of higher level cognitive skills such as language development, problem solving, inference generating, and critical thinking (Schunk, 2015).

In the context of computer aided learning, behaviorism has inspired many of the best known educational technologies, from early computer-assisted instruction (CAI) to present-day page-turners and drill-and-practice games (Ertmer & Newby, 2013; Peters, 2014). The specific assumptions or principles for the usage of behaviorist theory to instructional design can be referred to Table 3. One of the major limitation for CAI grounded on the behaviorist theory is that the learning material is often presented in small, isolated chunks of knowledge which is less to emphasize on connecting the pieces (Peters, 2014).

Cognitivism

Cognitivism as a learning theory can be traced back to the early twentieth century when psychologists and educators began to emphasize more on the complex cognitive processes such as thinking, problem solving, decision making, language, concept formation and information processing (Ertmer & Newby, 2013; Peters, 2014; Yilmaz, 2011). From the learning perspective, Cognitive theories focus on the conceptualization of students' learning processes and address the issues of how information is received, organized, stored, and retrieved by the mind (Arshavskiy, 2013; Ertmer & Newby, 2013).

Table 3. Specific principles (behaviourism) to instructional design (Adopted from Ertmer & Newby, 2013)

Learning Theory	Principles to instructional design (ID)	Possible ID applications
Behaviourism	An emphasis on producing observable and measurable outcomes in students	behavioural objectives, task analysis, criterion-referenced assessment
	Pre-assessment of students to determine where instruction should begin	learner analysis
	Emphasis on mastering early steps before progressing to more complex levels of performance	sequencing of instructional presentation, mastery learning
	Use of reinforcement to impact performance	tangible rewards, informative feedback
	Use of cues, shaping and practice to ensure a strong stimulus-response association	simple to complex sequencing of practice, use of prompts

Cognitivists assert that we learn better when we can connect new information to things we already know (Peters, 2014). Thus, learning is concerned not so much with what learners do but with what they know and how they come to acquire it (Jonassen, 1991). The learner is viewed as a very active participant in the learning process that involved cognitive activities (Arshavskiy, 2013).

Cognitivism has inspired the educational technology especially for the design and development of the intelligent tutoring system (ITS). Intelligent tutoring system is defined as a computer system that aims to provide immediate and customized instruction or feedback to the learners (Almurshidi & Naser, 2017). ITS adapts to an individual student's performance automatically by drawing on the knowledge incorporated into its database instead of just the predetermined questions, answers, and predefined pathways that made up behaviorist's CAI technologies (Peters, 2014). The specific assumptions or principles for the usage of cognitivist theory to instructional design can be referred to Table 4.

Cognitive Load Theory and Multimedia Learning

Cognitive load theory was proposed by John Sweller in the late 1980s (Sweller, 1988). The basic idea for cognitive load theory is that the cognitive capacity in working memory is limited and thus learning tasks should be designed in a way to avoid cognitive overload (Jong, 2010). Extensive research studies

Table 4. Specific principles (cognitivism) to instructional design (Adopted from Ertmer & Newby, 2013)

Learning Theory	Principles to instructional design (ID)	Possible ID applications
Cognitivism	Emphasis on the active involvement of the learner in the learning process	learner control, metacognitive training (e.g., self-planning, monitoring, and revising techniques)
	Use of hierarchical analyses to identify and illustrate prerequisite relationships	cognitive task analysis procedures
	Emphasis on structuring, organizing, and sequencing information to facilitate optimal processing	use of cognitive strategies such as outlining, summaries, synthesizers, advance organizers, etc.
	Creation of learning environments that allow and encourage students to make connections with previously learned material	recall of prerequisite skill; use of relevant examples, analogies

have been conducted grounded on cognitive load theory to facilitate learning process by optimizing the working memory capacity (Peters, 2014). Based on the cognitive load theory and findings from cognitive science, Mayer and his team developed the cognitive theory of multimedia learning which specifically deals with the multimedia design issues for education (Clark & Mayer, 2011). The cognitive theory of multimedia learning is based on three assumptions as follows:

1. We process visual and auditory information through separate channels (dual channel processing)
2. We are limited in the amount of information we take into either channel at once.
3. When we are engaged in active learning, we are not passively receiving information. Instead, we a) pay attention, b) organize incoming information (picking and choosing what's important), and c) integrate incoming information with other knowledge. We do all this in order to build a mental model of the key parts and relationships of the information we are presented with.

Based on the three assumptions, Mayer and his research team proposed the guideline for multimedia design principles supported by strong empirical research findings. According to Peters (2014), Mayer's research in the past fifteen years has led to the development of a number of research-based

Table 5. Mayer's principles of multimedia learning

Mayer's Principles of Multimedia Learning	
Multimedia principle	Segmenting principle
Contiguity principle	Pre-training principle
Modality principle	Signaling principle
Coherence principle	Voice principle
Personalization principle	Image principle
Redundancy principle	Individual differences principle

multimedia learning design principles, many of which pertain specifically to interface design. The Mayer's principles of multimedia learning can be referred to Table 5.

Constructivism

Constructivism is a general learning theory, by which it emphasizes that people actively construct their own understanding of the world through interactive experiences. (Oviatt, 2013; Peters, 2014). As opposed to behaviorist and cognitivist, constructivist argue that knowledge is a function of how the individual creates meaning from his or her own experiences (Ertmer & Newby, 2013). Learning is not viewed as the acquisition and accumulation of a finite set of skills and facts (Tam, 2009). As further elaborated by Peters (2014):

"Knowledge, rather than being an objective matchup with reality, is an individual's interpretation and construction based on a unique collection of past experiences, prior knowledge, and ways of interpreting things."

Constructivism has multiple roots in the philosophical and psychological viewpoints of this century, specifically in the works of Piaget, Bruner, and Goodman (Perkins, 1991). Constructivism has received increased attention in a number of different disciplines, including instructional design (Bednar et al., 1991; Ertmer & Newby, 2013; Ractham et al., 2012; Schrader, 2015). In the educational technology context, constructivist technology might provide tools for group discussion and knowledge building like wikis, collaborative media-making tools, discussion forums or chat rooms. (Peters, 2014). In addition, the web based 3-D worlds that let learners engage in virtual fieldwork experiences, explore virtual environments, making hypotheses, collect various

Table 6. Specific principles (constructivism) to instructional design (Adopted from Ertmer & Newby, 2013)

Learning Theory	Principles to instructional design (ID)	Possible ID applications
Constructivism	An emphasis on the identification of the context in which the skills will be learned and subsequently applied	anchoring learning in meaningful contexts
	An emphasis on learner control and the capability of the learner to manipulate information	actively using what is learned
	The need for information to be presented in a variety of different way	revisiting content at different times, in rearranged contexts, for different purposes, and from different conceptual perspectives
	Supporting the use of problem solving skills that allow leaners to go "beyond the information given"	developing pattern-recognition skills, presenting alternative ways of representing problems
	Assessment focused on transfer of knowledge and skills	presenting new problems and situations that differ from the conditions of the initial instruction

types of data, and propose solutions draw on constructivist learning theory as well. (Peters, 2014). However, Tam (2009) raised the concern that computer-supported constructivist environments should not only limit to involve the knowledge and intelligence to guide and structure the learning process, but rather should create situations and offer tools that stimulate students to maximize the use of their cognitive potential. The specific assumptions or principles for the usage of constructivist theory to instructional design can be refer to Table 6.

Connectivism

Connectivism is a relatively new learning theory as compared to behaviorism, cognitivism and constructivism. It is a learning theory promoted by Stephen Downes and George Siemens in the mid of 2000s that takes into account the needs and realities of the knowledge age. (Peters, 2014). In connectivism, the starting point for learning occurs when knowledge is actuated through the process of a learner connecting to and feeding information into a learning community (Goldie, 2016; Kop & Hill, 2008). As defined by Siemens (2005), "A community is the clustering of similar areas of interest that allows for interaction, sharing, dialoguing, and thinking together." Siemens (2006) further argued that a new learning theory, in fact, is required, due to the exponential growth and complexity of information available on the Internet, new possibilities for people to communicate on global networks, and for the ability to aggregate different information streams. In general, connectivism's principles are shifting focus from content itself to the connections of content (AlDahdouh, 2017). Similarly, our ability to connect to new knowledge is more important than how much we actually already know. (Peters, 2014).

Duke, Harper, & Johnston (2013) further summarized that connectivism could be a learning theory based on the following three reasons:

1. Connectivism is characterized as the enhancement of how a student learns with the knowledge and perception gained through the addition of a personal network. It is only through these personal networks that the learner can acquire the viewpoint and diversity of opinion to learn to make critical decisions through collaboration.
2. The sheer amount of data available makes it impossible for a learner to know all that is needed to critically examine a specific situation. Being able to tap into huge databases of knowledge in an instant empowers a learner to seek further knowledge.
3. Explaining learning by means of traditional learning theories is severely limited by the rapid change brought about by technology. Connectivism is defined as actionable knowledge, where an understanding of where to find knowledge may be more important than answering how or what that knowledge encompasses.

In contrast, some of the researchers argued on the validity of connectivism as a learning theory (Duke et al., 2013; Kerr, 2006; Weber & Vas, 2016). As described by Duke et al (2013), connectivism is a pedagogical view rather than a learning theory since the learning theories should address the issue of how to enable the learner at the instructional level not just remain at the curriculum level (focus on what is learned and why). This is consistent with the argument by Kerr (2006) that connectivism is claimed to be part of the existing learning theories, where various technologies only affect methods of instruction in numerous ways.

While the debate on connectivism theory will continue for many years, connectivism has become an important school of thought applicable to the use of technology to support for classroom learning such as the online learning with the combination of 3D interactive graphics, artificial neural networks and web technologies. (AlDahdouh, 2017; Kop & Hill, 2008; Peters, 2014).

Table 7. Descriptions, advantages and disadvantages of the five primary interaction styles (Adopted from Stone et al., 2005)

Interaction Style	Descriptions	Advantages	Disadvantages
Command Line	The command line interface was the first interactive dialog style to be commonly used. It provides a means of directly instructing the system, using function keys on a keyboard, single characters, abbreviations, or whole-word commands.	■ Is versatile and flexible ■ Appeals to expert users ■ Supports users' initiative by allowing them to define macros and shortcuts	■ Requires substantial training and memorization of commands
Menu Selection	A menu is a set of options from which the user must choose. Typically, the interface displays the options as menu items or icons and the user indicates a choice with a pointing device or keystroke, receiving feedback that indicates which option he or she has chosen, and the outcome of the command being executed.	■ Is easy to learn ■ Involves fewer keystrokes than command line ■ Structures decision making by breaking down the functionality into a set of menu items ■ Is good for learners and infrequent users	■ Presents the danger of creating too many menus and complex menu hierarchies ■ May slow frequent users who would prefer to use commands or shortcuts ■ Consumes screen space
Form-fill	Form-fill interfaces allow for easy movement around the form and for some fields to be left blank. The users work through the form, entering appropriate values.	■ Simplifies data entry ■ May require modest training ■ Assists users by providing defaults	■ Consumes screen space
Direct Manipulation	Direct manipulation (DM) interfaces allow user to interact directly with the user interface objects. In DM interfaces, the keyboard entry of commands or menu choices is replaced by manipulating a visible set of objects and actions.	■ Presents the task concepts visually – the user can see the task objects and act on them directly ■ Is easy to learn ■ Is easy to remember how to use ■ Avoids errors and allows easy recovery from errors if they occur ■ Encourage exploration	■ Requires graphic displays and continuous input devices ■ Presents the danger that icons and metaphors may have different meanings for different user groups.
Anthropomorphic interfaces	Anthropomorphic interface aim to interact with users in the same way that humans interact with each other. Natural language interfaces and interfaces that recognize gestures, facial expressions, or eye movements all belong to this category.	■ Can relieve the burden of learning the syntax for the interaction with the system.	■ Can be unpredictable ■ Difficult to implement

Table 8. Interaction types and technology involved (Adopted from Evans & Sabry, 2003)

<table>
<tr><th>No.</th><th colspan="2">Interaction Types (Learning Perspective)</th><th>The computer technology involved</th></tr>
<tr><td>1</td><td colspan="2">Student-student interaction</td><td rowspan="2">Electronic discussions boards, text chat rooms, video conferencing or electronic white-boarding.</td></tr>
<tr><td>2</td><td colspan="2">Teacher-student interaction</td></tr>
<tr><td rowspan="2">3</td><td colspan="2">Student-content interaction</td><td rowspan="2">Computer-based learning packages such as multimedia CD-ROMs and the Web.</td></tr>
<tr><td>Student-initiated interaction</td><td>Computer-initiated interaction (Schar & Krueger, 2000)</td></tr>
</table>

Interaction Styles

Interaction styles are the different ways that a user can communicate with a computer system and vice versa. As defined by Stone et al, (2005), "*an interaction style is a collection of user interface controls and their associated behavior. The interaction style provides both the look (appearance) and feel (behavior) of the user interface components indicating the way a user will communicate with the system.*" There are several interaction styles from which a designer may choose and the details are listed in Table 7.

TYPES OF INTERACTIONS

There are three types of interactions that shape the learning process (Moore 2012; More 1989): learner-content interaction, learner-instructor interaction, and learner-learner interaction.

1. **Learner-Content Interaction**: The first type of interaction is interaction between the learner and the content or subject of study. This is a defining characteristic of education. Without it there cannot be education, since it is the process of intellectually interacting with content that results in changes in the learner's understanding, the learner's perspective, or the cognitive structures of the learner's mind.
2. **Learner-Instructor Interaction**: The second type of interaction is regarded as essential by many educators, and as highly desirable by many learners is the interaction between the learner and the expert who prepared the subject material, or some other expert acting as instructor.

Figure 1. The three-way model applied to computer-initiated interactivity (Evans & Sabry, 2003)

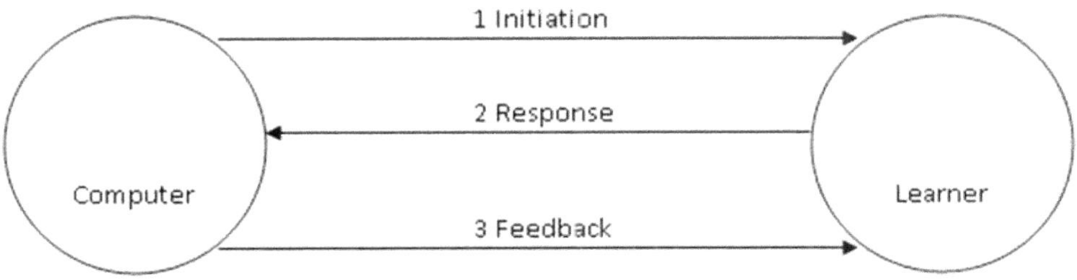

3. **Learner-Learner Interaction**: It is the third form of interaction, a new dimension of distance education that will be a challenge to our thinking and practice. This is inter-learner interaction, between one learner and other learners, alone or in group settings, with or without the real-time presence of an instructor.

The three types of interactions as proposed by Moore (1989) were further discussed by Evan and Sabry (2003) by which each of these types of interaction may make use of the computer technology and the related works are summarized in Table 8.

In Table 8, the student-content interaction can be further categorized into student-initiated interaction and computer-initiated interaction (Schar & Krueger, 2000). From the computer-initiated interaction perspective, Evans and Sabry (2003) proposed and formulated an interaction model called "The Three-Way Model of Interactivity (3-WMI)".

This model consists of three sequential actions (as refer to Figure 1):

1. Computer initiation: the computer presents a task or a series of options to the learner, in the form of a question or button (e.g., click here to begin);
2. Learner response: the learner selects an action in response to the options presented by the computer in the initiation stage;
3. Computer feedback: the computer presents the learner a new screen containing an assessment of the learner's response.

Figure 2. Illustration of the interactivity cycle (Evan & Sabry, 2003)

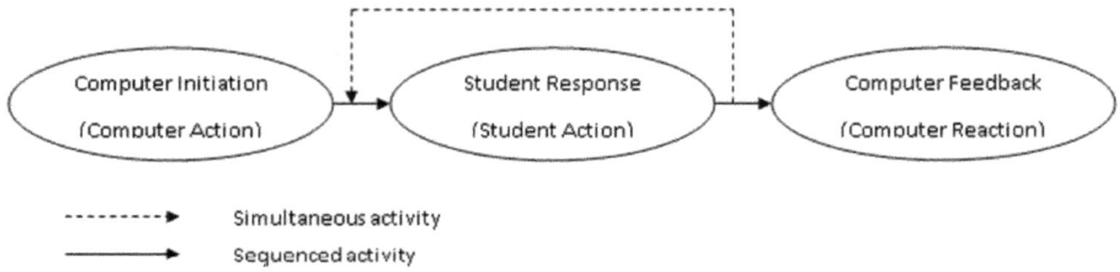

In addition, the three actions of the three-way model may form part of an iterative interactivity cycle. In such a cycle, computer feedback simultaneously initiates another interaction as illustrated in Figure 2. The only requirement for computer-initiated activity cycle is that they must begin with computer initiation and ultimately terminate with computer feedback.

Sometimes, the variety of forms of interaction often creates confusion over the meaning of the use of the word interactivity. In order to clarify what Evans and Gibbons (2007) meant by the term interactive system, they adopted the 3-way interaction model as formulated by Evans and Sabry (2003). Hillman et al. (1994) added a fourth component on learner-interface interaction to the literature discussion. They defined learner-interface interaction as "*a process of manipulating tools to accomplish a tas*k". They stressed the importance of learner-interface interaction because the learner must interact with the technological medium in order to interact with the content, instructor, or other learners.

A good interface design that may facilitate the learning will follow certain usability benchmark as the design consideration and reference. The usability benchmark and standards proposed for interface designs will be discussed in the next section.

USABILITY BENCHMARK AND STANDARDS

Standards are impartial best practices or definitions that act at as a national, multinational or international level (Brooks, 2015). As defined in the Oxford dictionary, "standard is a level of quality or attainment or something used as a measure, norm, or model in comparative evaluations (Oxford, 2017). According to Bevan (2006), Human Computer Interactions (HCI) standards have been developed over the last 20 years. International standards for HCI are developed

Table 9. International standards related to HCI context (adapted from Bevan, 2006)

Section	Principles and recommendations
1) Context and test methods	ISO/IEC 9126-1: Software Engineering – Product quality – Quality model
	ISO/ IEC TR 9126-4: Software Engineering – Product quality – Quality in use metrics
	ISO 9241-11: Guidance on Usability
	ISO/IEC DTR 19764 Guidelines methodology, and reference criteria for cultural and linguistic adaptability in information technology products
2) Software interface and interaction	ISO/IEC TR 9126-2: Software Engineering – Product quality – External metrics
	ISO/IEC TR 9126-3: Software Engineering – Product quality – Internal metrics
	ISO 9241: Ergonomic requirements for office work with visual display terminals. Part 10-17
	ISO 14915: Software ergonomics for multimedia user interfaces
	ISO TS 16071: Software accessibility
	ISO TR 19765 Survey of existing icons and symbols for elderly and disabled persons
	ISO TR 19766 Design requirements for icons and symbols for elderly and disabled persons
	ISO CD 23974: Software ergonomics for World Wide Web user interfaces
	IEC TR 61997: Guidelines for the user interfaces in multimedia equipment for general purpose use
3) Hardware interface	ISO 11064: Ergonomic design of control centres
	ISO/IEC TR15440 Future keyboards and other associated input devices and related entry methods
4) Development process	ISO 13407: Human-centred design processes for interactive systems
	ISO TR 16982: Usability methods supporting human centred design
5) Usability capability	ISO TR 18529: Human-centred lifecycle process descriptions
	ISO PAS 18152: A specification for the process assessment of human-system issues
6) Other related standards	ISO 9241-1: General Introduction
	ISO 9241-2: Guidance on task requirements
	ISO 10075-1: Ergonomic principles related to mental workload – General terms and definitions

under the auspices of the International Organisation for Standardisation (ISO) and the International Electrotechnical Commission (IEC). Standards related to usability can be categorized as primarily concerned with (Bevan, 2006):

Table 10. Usability Factors and its Descriptions according to ISO 9241-11 (ISO, 1998)

Usability factors	Descriptions
Effectiveness	The accuracy and completeness with which users achieve specified goals.
Efficiency	The resources expended in relation to the accuracy and completeness with which users achieve goals.
Satisfaction	The comfort and acceptability of use.

1. The use of the product (effectiveness, efficiency and satisfaction in a particular context of use).
2. The user interface and interaction.
3. The process used to develop the product.
4. The capability of an organisation to apply user centred design.

The international standards related to HCI context are listed in Table 9.

ISO 9241-11 Standard: Guidance on Usability (1998)

ISO 9241-11 standard provides the guidance on usability design by which it refers to the extent to which a product is usable. As defined in ISO 9241-11 standard, "*usability refer to the extent to which a product can be used by specified users to achieve specified goals with effectiveness, efficiency and satisfaction in a specified context of use*" (ISO, 1998). ISO documentation further provides the definitions regarding what is mean by effectiveness, efficiency and satisfaction as shown in Table 10.

As also described by Bevan et al. (2015), effectiveness means success in achieving goals, efficiency means not wasting time and satisfaction means willingness to use the system.

ISO 9241-210: Human-Centred Design for Interactive Systems (2010)

The ISO 9241-210 standard provides a framework for human-centred design (HCD) activities comprising the four stages: context of use, specification of user and organizational requirements, design solutions, and evaluation of design against requirements (ISO, 2010). These activities are carried out in an iterative fashion with the cycle being repeated until the design objectives have been attained (Maguire, 2013). According to Bevan (2014), ISO 9241-210

Table 11. Rationale for adopting human centered design principle (adapted from Bevan, 2014)

	Rationale for adopting principle of human centred design
1	The design is based upon an explicit understanding of users, tasks and environments
2	Users are involved throughout design and development
3	The design is driven and refined by user-centred evaluation
4	The process is iterative
5	The design addresses the whole user experience
6	The design team includes multi-disciplinary skills and perspectives

uses the phrase "human-centred" rather than "user-centred" to acknowledge the importance of stakeholders who may not be users. As further claimed by Bevan (2014), ISO 9241-210 standard provides one of the best concise introductions to usability for designers' reference. There are many supportive reasons for adopting the human centred design principle as shown in Table 11.

Justification on the Standards Reference

Throughout this research study especially on the design and development of the enhanced multimedia TAPS package, the ISO 9241-11 and ISO 9241-210 standards were adopted as reference guidelines for the usability designs. This was due to the following reasons:

1. The best practices stated in the standards can be a guideline for the usability design since the standards were contributed from the industry experts with the accumulation of vast experience (all these years). As raised by Brooks (2015), it is possible to instantly leverage the experience of industry experts and adopt the best practices if any designer is familiar with the standards.
2. By following the international standard as a guideline can provide authoritative evidence to cite regarding the user interface design decision. This is parallel with the statement that the standards have been highly successful in providing an internationally accepted basis for understanding and applying usability based on human centred design (Bevan et al., 2015; Maguire. 2013).

Table 12. Quality model for software with usability factors (Adopted from Kabir et al., 2016)

	Published Year	Quality Model Name	Usability Factors	Source
1	1977	McCall	Operability, Training, Communicativeness	(McCall et al., 1977)
2	1978	Boehm	Reliability, Efficiency, Human Engineering	(Boehm, 1978)
3	1991	Shackel	Effectiveness, Learnability, Flexibility, Attitude	(Shackel, 1991)
4	1992	FURPS	Human Factors, Aesthetics, Consistency, Documentation, Responsiveness	(Grady, 1992)
5	1993	Nielsen	Learnability, Efficiency, Memorability, Errors, Satisfaction	(Nielsen, 1993)
6	1998	SUMI	Efficiency, Affect, Helpfulness, Control, Learnability	Kirakowski & Corbett (1993)
7	1998	ISO 9242-11	Effectiveness, Efficiency, Satisfaction	(ISO, 1998)
8	2001	ISO 9126	Understandability, Learnability, Operability, Attractiveness, Usability Compliance	(ISO, 2001)
9	2006	QUIM	Productivity, Efficiency, Effectiveness, Safety, Learnability, Accessibility Satisfaction, Truthfulness, Universality, Usefulness	(Padda, 2009)
10	2014	SEM	Understandability, Learnability, Applicability, Effectiveness / Usefulness for Future Projects, User Satisfaction.	(Masood et al., 2014)

However, it should be noted that the ISO 9241-11 and ISO 9241-210 contents provide valuable direction as a reference rather than tell the designer on how to design (Brooke, 2015).

Usability Testing and Its Instruments

Usability is one of the important quality attributes to measure how usable and effective/ useful is the software system (Masood et al., 2014). According to Nielsen (2012), usability is one of the quality attribute that assesses how easy user interfaces are to use. Nielsen further add on that there are five quality components used to define usability which are learnability, efficiency, memorability, errors and satisfaction. Concept of usability has been widely applied to improve the quality of software systems, human computer interaction, software process improvement, and software process descriptions (Masood et al., 2014). In the context of software engineering, many researchers have contributed the quality models to measure and assure the software quality through various quality factors such as usability, learnability and efficiency.

Table 13. Advantages of standardized usability questionnaires (adopted from Nunnally, 1978)

Advantages	Descriptions
Objectivity	A standardized measurement supports objectivity because it allows usability practitioners to independently verify the measurement statements of other practitioners.
Replicability	It is easier to replicate the studies of others, or even one's own studies, when using standardized methods.
Quantification	Standardized measurements allow practitioners to report results in finer detail than they could using only personal judgement. In addition, standardization also permits practitioners to use powerful methods of mathematics and statistics to better understand their results.
Economy	Developing standardized measures requires a substantial amount of work. However, once developed, they are very economical to reuse.
Communication	It is easier for practitioners to communicate effectively when standardized measures are available. Inadequate efficiency and fidelity of communication in any field impedes progress.
Scientific generalization	Scientific generalization is at the heart of scientific work. Standardization is essential for accessing the generalization of results.

(Kabir et al., 2016). Among all these factors, usability is the major quality factor due to its impact on software acceptance (Kabir et al., 2016). The various quality model for software with usability factors can be referred to Table 12.

In the next section, the standardize usability instruments will be discussed including the specific usability questionnaires that were used in this research study.

Standardized Usability Questionnaires

A standardized questionnaire is a questionnaire designed for repeated use, typically with a specific set of questions presented in a specified order using a specified format, with specific rules for producing metrics based on the answers of respondents (Sauro & Lewis, 2012). As part of the development of standardized questionnaires, it is customary for the developer to report measurements of its reliability, validity and sensitivity, in other words, for the questionnaire to have undergone psychometric qualification (Nunnally, 1978). The standardized usability questionnaires can be further defined as the standardized questionnaires designed to assess participants' satisfaction with the perceived usability of products or systems during or immediately after usability testing (Sauro & Lewis, 2012; Garcia, 2013). Standardized

Table 14. Common and widely used standardized usability questionnaires

	Specific Standardized Usability Questionnaire	Source
1	Questionnaire for User Interaction Satisfaction (QUIS)	(Chin et al., 1988)
2	Software Usability Measurement Inventory (SUMI)	(Kirakowski & Corbett, 1993; McSweeney, 1992)
3	Post-Study System Usability Questionnaire (PSSUQ)	(Lewis 1990a, 1992, 1995, 2002)
4	Software Usability Scale (SUS)	(Brooke, 1996)
5	After-Scenario Questionnaire (ASQ)	(Lewis, 1990b, 1991, 1995)
6	Expectation ratings (ER)	(Albert & Dixon, 2003)
7	Usability Magnitude Estimation (UME)	(McGee, 2003, 2004)
8	Single Ease Question (SEQ)	(Sauro, 2010; Tedesco & Tullis, 2006)
9	Subjective Mental Effort Question (SMEQ)	(Sauro and Dumas, 2009)

measures offer many advantages to practitioners as summarized and shown in Table 13.

The common and widely used standardize usability questionnaires and its source are summarized as shown in Table 14. Some of the standardized usability questionnaires required the licensing fees for using the instruments for usability testing such as Questionnaire for User Interaction Satisfaction (QUIS) and Software Usability Measurement Inventory (SUMI). In contrast, some of the standardized usability questionnaires are available for free such as the Post-Study System Usability Questionnaire (PSSUQ) and Software Usability Scale (SUS).

System Usability Scale (SUS)

SUS usability evaluation instrument was developed by John Brooke in 1986 (Brooke, 1996). It is a 10 questions long survey instrument, each with five scale steps. The odd-numbered items have a positive tone while the even-numbered items have a negative tone. SUS can be used to quantify the usability of websites, applications, or any software or hardware that users interact with (Rotolo, 2017). The SUS questionnaire is highly reliable (Cronbach's alpha value = 0.91) and free of charges, thus widely accepted as one of the popular standardized usability questionnaire in the market (Garcia, 2013). It is also widely used in the context of educational software for the usability testing. Table 15 compiled with the list on the usage of SUS for educational software as found in the literature.

Table 15. Compilation of the SUS usage for educational software evaluation

Educational Software	Purpose	Source
e-learning platform evaluation	Their aim was to detect usability problems and to cultivate knowledge of web developers on their end-users and user centred design.	(Renaut, Christophe, Flory, & Heyde, 2006)
DELTA (a distributed learning resources repository)	They used SUS to evaluate the perceived usability of DELTA.	(Venturi & Bessis, 2006)
Edutainment platforms (Virtual Classroom, Game-based & Storytelling).	Their purpose was to evaluate three edutainment platforms in terms of user performance, learning effectiveness and satisfaction in order to explore usability aspect of educational entertainment in e-Learning.	(Ayad & Rigas, 2010)
UNITE (an e-learning platform)	SUS was used as one of the usability evaluation technique to evaluate UNITE, an e-learning platform for secondary schools.	(Granić & Ćukušić, 2011)
Virtual learning environment	They used SUS to evaluate the usability of the virtual learning environment adopted by the Distance Education Center of the Federal Institute of Espírito Santo – Brazil, which has the Moodle platform as a basis.	(Simões & de Moraes, 2012)
Topolor system	Their aim was to assess the usability of the first version of the Topolor system a Social Personalized Adaptive E-Learning Environment (SPAEE)	(Shi, Awan, & Cristea, 2013)
Moodle system with the Drag & Share tool	Their aim was to evaluate the user satisfaction for the Moddle system with and without the Drag & Share tool.	(Marco, Penichet, & Gallud, 2013)
Simulation-based learning system	Their aim was to assess the simulation-based learning system perceived usability.	(Luo, Liu, Kuo, & Yuan, 2014)
Moodle-based Learning Management Systems (LMSs)	SUS was used to evaluate the perceived usability of LMSs.	(Orfanou, Tselios, & Katsanos, 2015)
MyGrammerTest (myGraTe) mobile application	SUS was used to evaluate the perceived usability of the respective mobile application.	(Ganapathy, Shuib, & Azizan, 2016)

According to Brooke (1996), participants should complete the SUS questionnaire after having used the system under evaluation but before any debriefing or other discussion. The SUS scoring method requires participants to provide a response to all 10 items. Based on the descriptions by Sauro and Lewis (2012), the very first step in scoring a SUS is to determine the contribution of each item's score, which range from 0 to 4. For positively worded items (odd numbers), the score contribution is the scale position minus 1 (xi-1). For negatively worded items (even numbers), the score contribution is 5 minus the scale position (5- xi). To get the overall SUS score, the sum of the item score contributions need to be multiplied by 2.5. Thus, overall SUS scores range from 0 to 100 in 2.5-point increments.

Table 16. Basic statistical information for SUS

Statistic	Bangor et al. (2008)	Lewis and Sauro (2009)		
	Overall	Overall	Usable	Learnable
N	2,324	324	324	324
Minimum	0	7.5	0	0
Maximum	100	100	100	100
Mean	70.14	62.1	59.44	72.72
Variance	471.32	494.38	531.54	674.47
Standard deviation	21.71	22.24	23.06	25.97
Standard error of the mean	0.45	1.24	1.28	1.44
Skewness	NA	-0.43	-0.38	-0.8
Kurtosis	NA	-0.61	-0.6	-0.17
First quartile	55	45	40.63	50
Median	75	65	62.5	75
Third quartile	87.5	75	78.13	100
Interquartile range	32.5	30	37.5	50
Critical z (99.9%)	3.09	3.09	3.09	3.09
Critical d (99.9%)	1.39	3.82	3.96	4.46
99.9% confidence interval upper limit	71.53	65.92	63.4	77.18
99.9% confidence interval lower limit	68.75	58.28	55.48	68.27
Note: Add and subtract critical d (computed by multiplying the critical z and the standard error) from the mean to get the upper and lower bounds of the 99.9% confidence interval.				

Through the literature search, recent research on the psychometric properties of the SUS has provided some normative data for reference. The basic statistical information about the SUS from the data reported by Bangor et al. (2008) and Lewis and Sauro (2009) can be referred to Table 16.

Another significant contribution was by Sauro (2011). He analysed the data from 3,187 completed SUS questionnaires and derived the curved grading scale interpretation of SUS scores as shown in Table 17. This curved grading scale interpretation table contributes as a norm table for SUS scores.

Table 17. Curved grading scale interpretation of SUS scores (Sauro, 2011)

SUS Score Range	Grade	Percentile Range
84.1-100	A+	96-100
80.8-84	A	90-95
78.9-80.7	A-	85-89
77.2-78.8	B+	80-84
74.1-77.1	B	70-79
72.6-74	B-	65-69
71.1-72.5	C+	60-64
65-71	C	41-59
62.7-64.9	C-	35-40
51.7-62.6	D	15-34
0-51.7	F	0-14

Post-Study System Usability Questionnaire (PSSUQ)

The PSSUQ is a questionnaire designed to assess users' perceived satisfaction with computer systems or applications (Sauro & Lewis, 2012). PSSUQ questionnaire is highly reliable (Cronbach's alpha value = 0.94) and free of charges (Garcia, 2013). The latest third version of PSSUQ has 16-items in the questionnaire.

The PSSUQ items produce four scores – one overall and three subscales. The rules for computing them are:

- Overall: Average responses for Item 1 through 16 (all the items)
- System Quality(SysQual): Average Items 1 through 6
- Information Quality (InfoQual): Average Items 7 through 12
- Interface Quality (IntQual): Average Items 13 through 15

According to Sauro and Lewis (2012), the resulting scores for each item in PSSUQ questionnaire can take values between 1 and 7, with lower scores indicating a higher degree of satisfaction. The interpretation of the PSSUQ data is best performed through the comparison with PSSUQ Norms. The best available norms for PSSUQ version 3 (means and 99% confidence intervals) is shown in Table 18 (Sauro and Lewis, 2012).

Table 18. PSSUQ version 3 norms (Means and 99% Confidence Intervals)

Item	Item Text	Lower Limit	Mean	Upper Limit
1	Overall, I am satisfied with how easy it is to use this system.	2.6	2.85	3.09
2	It was simple to use this system.	2.45	2.69	2.93
3	I was able to complete the tasks and scenarios quickly using this system.	2.86	3.16	3.45
4	I felt comfortable using this system.	2.4	2.66	2.91
5	It was easy to learn to use this system.	2.07	2.27	2.48
6	I believe I could become productive quickly using this system.	2.54	2.86	3.17
7	The system gave error messages that clearly told me how to fix problems.	3.36	3.7	4.05
8	Whenever I made a mistake using the system, I could recover easily and quickly.	2.93	3.21	3.49
9	The information (e.g., online help, on-screen messages, and other documentation) provided with this system was clear.	2.65	2.96	3.27
10	It was easy to find the information I needed.	2.79	3.09	3.38
11	The information was effective in helping me complete the tasks and scenarios.	2.46	2.74	3.01
12	The organization of information on the system screens was clear.	2.41	2.66	2.92
13	The interface of this system was pleasant.	2.06	2.28	2.49
14	I liked using the interface of this system.	2.18	2.42	2.66
15	This system has all the functions and capabilities I expect it to have.	2.51	2.79	3.07
16	Overall, I am satisfied with this system.	2.55	2.82	3.09
Scale	Scale Scoring Rule			
SysUse	Average Items 1-6.	2.57	2.8	3.02
InfoQual	Average Items 7-12.	2.79	3.02	3.24
IntQual	Average Items 13-15.	2.28	2.49	2.71
Overall	Average Items 1-16.	2.62	2.82	3.02
Note: These data are from 21 studies and 210 participants, analysed at the participant level.				

Some experiments of this research adopted the SUS and PSSUQ instruments as the usability evaluation instruments to test the usability of the enhanced multimedia TAPS package due to the following reasons:

1. Bothe the SUS and PSSUQ usability questionnaires are available for free without any charges.
2. Both the SUS and PSSUQ usability questionnaires are widely accepted in the market with high reliability (both Cronbach's alpha value > 0.9).
3. The procedures for data analysis is straightforward with established norms to support data interpretations.

SUMMARY

In summary this Chapter discussed the issues on user interface design, user interface design in the computer aided learning environment and its challenge, learning theories for interface design and recommendations. Usability issues and benchmarks were also discussed, and various usability testing instruments that can be used for usability evaluation were identified.

REFERENCES

Albert, W., & Dixon, E. (2003). Is this what you expected? The use of expectation measures in usability testing *Usability Professionals Association Annual Conference*.

AlDahdouh, A. A. (2017). Does Artificial Neural Network support Connectivism's assumptions? *International Journal of Instructional Technology and Distance Learning*, *14*(3), 3–26.

Almurshidi, S. H., & Naser, S. S. A. (2017). Design and Development of Diabetes Intelligent Tutoring System. *European Academic Research*, *4*(9), 1–9.

Arshavskiy, M. (2013). Instructional Design for Elearning: Essential Guide to Creating Successful Elearning Courses. *Createspace Independent Publishing Platform*.

Atoum, I., & Bong, C. H. (2015). Measuring Software Quality in Use: State-of-the-Art and Research Challenges. *SQP*, *17*(2), 4–15.

Ayad, K., & Rigas, D. (2010). Comparing virtual classroom, game-based learning and storytelling teachings in e-learning. *International Journal of Education and Information Technologies*, *4*(1), 15–23.

Bangor, A., Kortum, P. T., & Miller, J. T. (2008). An Empirical Evaluation of the System Usability Scale. *International Journal of Human-Computer Interaction, 24*(6), 574–594. doi:10.1080/10447310802205776

Bednar, A. K., Cunningham, D., Duffy, T. M., & Perry, J. D. (1991). Theory into practice: How do we link? In G. J. Anglin (Ed.), *Instructional technology: Past, present, and future*. Englewood, CO: Libraries Unlimited.

Bevan, N. (2006). International Standards for HCI. In *Encyclopedia of Human Computer Interaction*. Idea Group Publishing. doi:10.4018/978-1-59140-562-7.ch056

Bevan, N. (2006). International Standards for HCI. In *Encyclopedia of Human Computer Interaction*. Idea Group Publishing. doi:10.4018/978-1-59140-562-7.ch056

Bevan, N. (2014). *How You Could Benefit from Using ISO Standards*. Academic Press.

Bevan, N., Carter, J., & Harker, S. (2015). ISO 9241-11 Revised: What Have We Learnt About Usability Since 1998? In *International Conference on Human-Computer Interaction 2015* (pp. 143–151). Springer Berlin.

Boehm, B. (1978). *Characteristics of Software Quality*. Amsterdam: North Holland Publishing.

Brooke, J. (1996). SUS- A quick and dirty usability scale. *Usability Evaluation in Industry, 189*(194), 4–7.

Brooks, P. (2015). *What on Earth is ISO 9241?* Retrieved April 16, 2017, from http://www.uxbooth.com/articles/what-on-earth-is-iso-9241/

Chao, G. (2009). Human-Computer Interaction: Process and Principles of Human-Computer Interface Design. In *2009 International Conference on Computer and Automation Engineering* (pp. 230–233). IEEE.

Chin, J. P., Diehl, V. A., Norman, K. L. (1988) *Development of an Instrument Measuring User Satisfaction of the Human-Computer Interface*. Academic Press.

Clark, R. C., & Mayer, R. E. (2011). *E-Learning and the Science of Instruction: Proven Guidelines for Consumers and Designers of Multimedia Learning*. San Francisco, CA: Pfeiffer. doi:10.1002/9781118255971 de Jong, T. (2010). Cognitive load theory, educational research, and instructional design: Some food for thought. *Instructional Science, 38*(2), 105–134. doi:10.100711251-009-9110-0

Duke, B., Harper, G., & Johnston, M. (2013). Connectivism as a Digital Age Learning Theory. *The International HETL Review*, 1–13.

Ertmer, P. A., & Newby, T. J. (2013). Behaviorism, Cognitivism, Constructivism: Comparing Critical Features from an Instructional Design Perspective. *Performance Improvement Quarterly, 26*(2), 43–71. doi:10.1002/piq.21143

Evans, C., & Sabry, K. (2003). Evaluation of the interactivity of web-based learning systems: Principles and proces. *Innovations in Education and Teaching International, 40*(1), 89–99. doi:10.1080/1355800032000038787

Ganapathy, M., Shuib, M., & Azizan, S. N. (2016). Malaysian ESL Students' Perceptions on the Usability of a Mobile Application for Grammar Test: A Case Study of ESL Undergraduates in Universiti Sains Malaysia. *The Southeast Asian Journal of English Language Studies, 22*(1), 127–140.

Garcia, A. (2013). *UX Research | Standardized Usability Questionnaire*. Retrieved April 18, 2017, from https://chaione.com/blog/ux-research-standardizingusability-questionnaires/

Goldie, S. G. J. (2016). Connectivism: A knowledge learning theory for the digital age? *Medical Teacher, 38*(10), 1064–1069. doi:10.3109/0142159X.2016.1173661 PMID:27128290

Grady, R. B. (1992). *Practical Software Metrics for Project Management and Process Improvement*. Prentice Hall.

Granić, A., & Ćukušić, M. (2011). Usability Testing and Expert Inspections Complemented by Educational Evaluation: A Case Study of an e-Learning Platform. *Journal of Educational Technology & Society, 14*(2), 107–123.

Hilgard, E. R., & Bower, G. H. (1966). *Theories of learning* (3rd ed.). Englewood Cliffs, NJ: Prentice Hall.

Hillman, D. C. A., Willis, D. J., & Gunawardena, C. N. (1994). Learner-interface interaction in distance education: An extension of contemporary models and strategies for practitioners. *American Journal of Distance Education, 8*(2), 30–42. doi:10.1080/08923649409526853

ISO. (1998). *ISO 9241-11 Ergonomic requirements for office work with visual display terminals (VDTs) — Part 11: Guidance on usability*. Retrieved April 16, 2017, from https://www.iso.org/obp/ui/#iso:std:iso:9241:-11:ed-1:v1:en

ISO. (2001). *ISO/IEC 9126-1:2001 Software engineering -- Product quality -- Part 1: Quality model*. Retrieved April 16, 2017, from https://www.iso.org/standard/22749.html

ISO. (2010). *ISO 9241-210 Ergonomics of human-system interaction — Part 210: Human-centred design for interactive systems*. Retrieved April 16, 2017, from https://www.iso.org/obp/ui/#iso:std:iso:9241:-210:ed-1:v1:en

Jonassen, D. H. (1991). Objectivism vs constructivism: Do we need a new philosophical paradigm. *Educational Technology Research and Development*, *39*(3), 5–14. doi:10.1007/BF02296434

Kabir, M. A., Rehman, M. U., & Majumdar, S. I. (2016). An analytical and comparative study of software usability quality factors. In *2016 7th IEEE International Conference on Software Engineering and Service Science (ICSESS)* (pp. 800–803). IEEE.

Kerr, B. (2006). *A challenge to connectivism*. Retrieved April 10, 2017, from http://billkerr2.blogspot.my/2006/12/challenge-to-connectivism.html

Kirakowski, J., & Corbett, M. (1993). SUMI: The Software Usability Measurement Inventory. *British Journal of Educational Technology*, *24*(3), 210–212. doi:10.1111/j.1467-8535.1993.tb00076.x

Kop, R., & Hill, A. (2008). Connectivism: Learning theory of the future or vestige of the past? *International Review of Research in Open and Distance Learning*, *9*(3), 1–13. doi:10.19173/irrodl.v9i3.523

Lewis, J. R. (1990a). Psychometric Evaluation of a Post-study System Usability Questionnaire: The PSSUQ. Boca Raton, FL: Academic Press.

Lewis, J. R. (1990b). Psychometric Evaluation of an After-Scenario Questionnaire for Computer Usability Studies: The ASQ. Boca Raton, FL: Academic Press. doi:10.1145/122672.122692

Lewis, J. R. (1991). Psychometric evaluation of an after-scenario questionnaire for computer usability studies. *ACM SIGCHI Bulletin, 23*(1), 78–81.

Lewis, J. R. (1992). Psychometric Evaluation of the Post-Study System Usability Questionnaire: The PSSUQ. *Proceedings of the Human Factors and Ergonomics Society Annual Meeting*, *36*(16), 1259–1260. doi:10.1177/154193129203601617

Lewis, J. R. (1995). IBM computer usability satisfaction questionnaires: Psychometric evaluation and instructions for use. *International Journal of Human-Computer Interaction*, *7*(1), 57–78. doi:10.1080/10447319509526110

Lewis, J. R., & Sauro, J. (2009). The Factor Structure of the System Usability Scale. In *Proceedings of the 1st International Conference on Human Centered Design* (pp. 94–103). Berlin: Springer-Verlag.

Luo, G.-H., Liu, E. Z.-F., Kuo, H.-W., & Yuan, S.-M. (2014). Design and implementation of a simulation-based learning system for international trade. *International Review of Research in Open and Distance Learning*, *15*(1), 203–226. doi:10.19173/irrodl.v15i1.1666

Luostarinen, R., Manner, J., Maatta, J., & Jarvinen, R. (2010). User-centered design of graphical user interfaces. In *2010 - MILCOM 2010 MILITARY COMMUNICATIONS CONFERENCE* (pp. 50–55). IEEE.

Maguire, M. (2013). Using Human Factors Standards to Support User Experience and Agile Design. Academic Press. doi:10.1007/978-3-642-39188-0_20

Manresa-Yee, C., Ponsa, P., Varona, J., & Perales, F. J. (2010). User experience to improve the usability of a vision-based interface. *Interacting with Computers*, *22*(6), 594–605.

Marco, F. A., Penichet, V. M. R., & Gallud, J. A. (2013). Collaborative e-Learning through Drag & Share in synchronous shared workspaces. *Journal of Universal Computer Science*, *19*(7), 894–911.

Masood, Z., Shang, X., & Yousaf, J. (2014). Usability Evaluation Framework for Software Engineering Methodologies. *Lecture Notes on Software Engineering*, *2*(3), 225–232.

McCall, J. A., Richards, P. K., & Walters, C. F. (1977). *Factors in Software Quality Concept and Definitions of Software Quality*. Academic Press.

McGee, M. (2003). Usability Magnitude Estimation. *Proceedings of the Human Factors and Ergonomics Society Annual Meeting*, *47*(4), 691–695. doi:10.1177/154193120304700406

McGee, M. (2004). Master usability scaling. In *Proceedings of the 2004 conference on Human factors in computing systems - CHI '04* (pp. 335–342). New York: ACM Press. doi:10.1145/985692.985735

McSweeney, R. (1992). *SUMI: A Psychometric Approach to Software Evaluation*. Cork, Ireland: University College of Cork.

Moore, M. G. (1989). Three types of interaction. *American Journal of Distance Education*, *3*(2), 1–6. doi:10.1080/08923648909526659

Moore, M. G. (2012). *Handbook of Distance Education Today* (3rd ed.). Routledge.

Nielsen, J. (1993). *Usability Engineering*. San Diego, CA: Morgan Kaufmann. doi:10.1016/B978-0-08-052029-2.50007-3

Nielsen, J. (2012). *Usability 101: Introduction to Usability*. Retrieved from https://www.nngroup.com/articles/usability-101-introduction-to-usability/

Norman, D. (2013). *The Design of Everyday Thing* (R. Exp, Ed.). Basic Books.

Nunnally, J. C. (1978). *Psychometric Theory*. New York: McGraw-Hill.

Obeidat, M. Z. A., & Salim, S. S. (2010). *Integrating user interface design guidelines with adaptation techniques to solve usability problems. In 2010 3rd International Conference on Advanced Computer Theory and Engineering (ICACTE) (pp. V1-280-V1-284)*. IEEE.

Orfanou, K., Tselios, N., & Katsanos, C. (2015). Perceived Usability Evaluation of Learning Management Systems: Empirical Evaluation of the System Usability Scale. *The International Review of Research in Open and Distributed Learning*, *16*(2), 57–65. doi:10.19173/irrodl.v16i2.1955

Oviatt, S. (2013). *The Design of Future Educational Interfaces* (1st ed.). Routledge. doi:10.4324/9780203366202

Oxford Dictionary. (2017). *English Oxford Living Dictionaries*. Retrieved August 7, 2017, from https://en.oxforddictionaries.com/definition

Padda, H. (2009). *QUIM: A Model for Usability/Quality in use Measurement*. LAP Lambert Academic Publishing.

Peters, D. (2014). *Interface Design for Learning: Design Strategies for Learning Experiences*. Pearson Education.

Quintana, C., Skin, N., Norris, C., & Soloway, E. (2006). Learner-Centred Design: Reflections on the Past and Directions for the Future. In R. K. Sawyer (Ed.), *The Cambridge Handbook of the Learning Sciences* (1st ed.; p. 648). London, UK: Cambridge University Press.

Ractham, P., Kaewkitipong, L., & Firpo, D. (2012). The Use of Facebook in an Introductory MIS Course: Social Constructivist Learning Environment*. *Decision Sciences Journal of Innovative Education*, *10*(2), 165–188. doi:10.1111/j.1540-4609.2011.00337.x

Renaut, C., Christophe, B., Flory, L., & Heyde, M. (2006). Improving web site usability for a better e-learning experience. *Current Developments in Technology-Assisted Education*, 891–895.

Rotolo, T. (2017). *SUS and PSSUQ: Quantifying user experiences*. Retrieved from https://www.trymyui.com/blog/2017/02/24/sus-pssuq-website-usability-surveys/

Sauro, J. (2010). *If you could only ask one question, use this one*. Retrieved April 14, 225 2017, from https://measuringu.com/single-question/

Sauro, J., & Dumas, J. S. (2009). Comparison of three one-question, post-task usability questionnaires. In *Proceedings of the 27th international conference on Human factors in computing systems - CHI 09* (p. 1599). New York: ACM Press. doi:10.1145/1518701.1518946

Sauro, J., & Lewis, J. R. (2012). *Quantifying The User Experience. Practical Statistics for User Research*. Morgan Kaufmann.

Sawyer, R. K. (2008). Optimizing Learning: Implications of Learning Sciences Research. In *Innovating to Learn, Learning to Innovate*. Paris, France: OECD. doi:10.1787/9789264047983-4-en

Schär, S., & Krueger, H. (2000). Using New Learning Technologies with Multimedia. *Journal of the Institute of Electrical and Electronics Engineers Multimedia*, *7*(3), 40–51.

Schrader, D. E. (2015). Constructivism and Learning in the Age of Social Media: Changing Minds and Learning Communities. *New Directions for Teaching and Learning, 2015*(144), 23–35.

Schunk, D. H. (2015). *Learning Theories: An Educational Perspective* (7th ed.). London, UK: Pearson Education Limited.

Shackel, B. (1991). Usability—context, framework, definition, design and evaluation. In *Human factors for informatics usability* (pp. 21–37). New York: Cambridge University Press.

Shi, L., Awan, M. S. K., & Cristea, A. I. (2013). Evaluating System Functionality in Social Personalized Adaptive E-Learning Systems. Academic Press. doi:10.1007/978-3-642-40814-4_87

Shneiderman, B., Plaisant, C., Cohen, M., Jacobs, S., Elmqvist, N., & Diakopoulos, N. (2017). *Designing the User Interface: Strategies for Effective Human-Computer Interaction* (6th ed.). New York: Pearson Education Limited.

Siemens, G. (2005). *Connectivism: Learning as Network-Creation*. Retrieved April 10, 2017, from http://www.elearnspace.org/Articles/networks.htm

Siemens, G. (2006). *Connectivism: Learning Theory or Pastime for the Self-Amused?* Retrieved April 10, 2017, from http://www.elearnspace.org/Articles/connectivism_self-amused.htm

Simandan, D. (2013). Introduction: Learning as a Geographical Process. *The Professional Geographer*, *65*(3), 363–368. doi:10.1080/00330124.2012.693872

Simões, A. P., & de Moraes, A. (2012). The ergonomic evaluation of a virtual learning environment usability. *Work (Reading, Mass.)*, *41*, 1140–1144. PMID:22316872

Stone, D., Jarrett, C., Woodroffe, M., & Minocha, S. (2005). *User interface design and evaluation*. San Francisco, CA: The Open University.

Sweller, J. (1988). Cognitive Load During Problem Solving: Effects on Learning. *Cognitive Science*, *12*(2), 257–285. doi:10.120715516709cog1202_4

Tam, M. (2009). Constructivism, Instructional Design, and Technology. In J. W. Willis (Ed.), *Constructivist Instructional Design (C-ID): Foundations, Models, and Examples* (pp. 61–80). Charlotte, NC: Information Age Publishing.

Tedesco, D. P., & Tullis, T. S. (2006). A comparison methods for eliciting post-task subjective ratings in usability testing. *Usability Professionals Association Annual Conference*.

Venturi, G., & Bessis, N. (2006). User-centred evaluation of an e-learning repository. In *Proceedings of the 4th Nordic conference on Human-computer interaction changing roles - NordiCHI'06* (pp. 203–211). New York: ACM Press. doi:10.1145/1182475.1182497

Weber, C., & Vas, R. (2016). Applying connectivism? Does the connectivity of concepts make a difference for learning and assessment? In *2016 IEEE International Conference on Systems, Man, and Cybernetics (SMC)* (pp. 2079–2084). IEEE.

Winn, W. (1990). Some implications of cognitive theory for instructional design. *Instructional Science*, *19*(1), 53–69. doi:10.1007/BF00377985

Woollard, J. (2011). *Psychology for the Classroom: Behaviourism* (1st ed.). New York: Routledge. doi:10.4324/9780203819166

Yilmaz, K. (2011). The Cognitive Perspective on Learning: Its Theoretical Underpinnings and Implications for Classroom Practices. *The Clearning House: A Journal of Educational Strategies, Issues and Ideas*, *84*(5), 204–212.

KEY TERMS AND DEFINITIONS

Behaviorism: Is a systematic approach to understanding the behavior of humans and other animals. It assumes that all behaviors are either reflexes produced by a response to certain stimuli in the environment, or a consequence of that individual's history, including especially reinforcement and punishment, together with the individual's current motivational state and controlling stimuli. Although behaviorists generally accept the important role of inheritance in determining behavior, they focus primarily on environmental factors.

Cognitivism: Cognitivism is the study in psychology that focuses on mental processes, including how people perceive, think, remember, learn, solve problems, and direct their attention to one stimulus rather than another.

Connectivism: Is a theory of learning in a digital age that emphasizes the role of social and cultural context in how and where learning occurs. Learning does not simply happen within an individual, but within and across the networks.

Constructivism: Is basically a theory based on observation and scientific study about how people learn. It says that people construct their own understanding and knowledge of the world, through experiencing things and reflecting on those experiences.

Post-Study System Usability Questionnaire (PSSUQ): Is another usability quantification survey, very similar to SUS, developed at the IBM Design Center in 1992. The survey: Like SUS, the survey is a series of statements describing the system, which users agree or disagree with using a Likert scale.

System Usability Scale (SUS): In systems engineering, the system usability scale (SUS) is a simple, ten-item attitude Likert scale giving a global view of subjective assessments of usability and system interactivity.

User Interface Design: Is the design of user interfaces for machines and software, such as computers, home appliances, mobile devices, and other electronic devices, with the focus on maximizing usability and the user experience.

Chapter 4
Computer Aided Simulation of a 4BL Problem With MATHLAB

ABSTRACT

In this chapter, the synthesis and analysis of four link bar (4BL) mechanisms problem, which is a well-known classical design problem, is discussed. However, limitations to the classical theory of this problem potentially limit its application to certain real-world problems by virtue of the small number of precision points and unspecified order. A prototype of the problem was designed and simulated for visualizing a four-bar link (4BL – simple movable closed chain linkage) mechanism engineering problem using MATLAB software. The aim is to facilitate the analysis, dynamic simulation of the four-bar mechanisms. First, a brief review of some of the current computer-aided learning software for four-bar mechanisms is presented. These software packages provide two-dimensional visualization and computational capabilities necessary to synthesize and analyze four-bar mechanisms. The chapter also reviews the kinematics of four-bar mechanisms as they pertain to their geometric modelling followed by the design approach of the graphical MATLAB simulation for 4BL mechanisms. A preliminary validation test using pre- and posttest questionnaires and focus group discussion with student participation in using the MATLAB simulation has facilitated in visualizing the engineering concepts of 4BL mechanism at UNITEN.

BACKGROUND

DOI: 10.4018/978-1-7998-0465-9.ch004

As mentioned in the previous Chapter, TAPS (Technology-Assisted Problem Solving) packages are some of the many multimedia courseware kinds that have many characteristics like better presentation and clarity, a step-by-step approach to solving problems, a user-friendly environment, impartial assessment of performance, and flexibility to incorporate 2-D and 3-D models. TAPS is involved with disciplines like Engineering, where visualization coupled with understanding cause and effects of concepts are the heart to mastering the material. The nature of TAPS provides students to make numerous attempts in a relatively short period of time and quickly see the consequences of their decision and strategies without the fear of failure or breaking of equipments. TAPS allows students to see the subject dynamics with all dimensions which helps to explain complex situations in a more comprehensive way. With this respect, it improves students' practical competencies; such as critical- thinking, decision-making and management skills.

MOTIVATION OF STUDY

This study focuses on the problems faced by students in Mechanical Engineering field, i.e. having difficulties in visualizing dynamic responses of mechanisms in UNITEN. A major challenge facing instructors in teaching the Engineering Mechanics Dynamics course at UNITEN is helping students relate the theory to the physical world. Past experiences in teaching the second year students indicated that there were students who found it difficult to visualize some of the concepts in Engineering and to apprehend theory and practical. In general, engineering textbooks cannot represent Mechanical actions such as movement of linkages, pistons, and crankshafts in the form of dynamic illustrations (animated forms). Since these Mechanical actions are very important in visualization experiments, the motions and actions of these objects maybe worthwhile to be shown in animated forms. Hence, the motivation is to implement a TAPS package for Mechanical Engineering students to better understand a sub-topic in dynamics. In this study the 4BL mechanism is selected based on the requirement of the instructor (expert domain). The objectives of the study were to study the problems faced by Mechanical Engineering student at UNITEN, implement a TAPS package to simulate and enhance Mechanical Engineering students' understanding on 4 -bar mechanism problem and to test and evaluate the TAPS package to find out if it could enhance the learning process.

BASIC THEORY OF 4BL MECHANISMS

One of the objectives of this research was to study and understand the motion of the 4BL mechanism and see how it could be simulated (Manjit, 2011). The purpose of this mechanism is to transmit motion and force from one location to another. The theory of this mechanism can be understood by the 4BL (which consists of four rigid bodies each attached to two others by single joints or pivots to form a closed loop) (Wikipedia, 2011) and (Oleg, 2000). This problem was chosen because in the past it was found that some students could not visualize its concepts in engineering at UNITEN. In addition we could not find any completed 2-D model of this problem in the present literatures. Some free available simulations found on the Internet (Fig. 1 and 2) however gave wrong results and misinterpreted the data and were not interactive enough. As such these simulations were not suitable and helpful for the students to learn.

To further explain, a 4BL or four-bar linkage is the simplest closed bar linkages that are important mechanisms found in machines. The kinematics and dynamics of 4BL are important topics in mechanical engineering. 4BL linkages are constructed from four links connected in a loop by four one degree of freedom joints. A joint may be either a revolute that is a hinged joint, denoted by ***R***, or a prismatic, as sliding joint, denoted by ***P***. The planar quadrilateral linkage is formed by four links and four revolute joints, denoted ***RRRR***. The slider-crank linkage is constructed from four links connected by three revolute and one prismatic joint, or ***RRRP***. The double slider is a ***PRRP*** linkage. One link of the chain is usually fixed, and is called the ground link, fixed link, or the frame. The two links connected to the frame are called the grounded links and are generally the input and output links of the system. The last link is the floating link, which is also called a coupler or connecting rod because it connects an input to the output.

The Grashof's condition for a planar quadrilateral linkage, which is a planar four-bar linkage constructed from four hinged joints, states: *If the sum of the shortest and longest link of a planar quadrilateral linkage is less than or equal to the sum of the remaining two links, then the shortest link can rotate fully with respect to a neighboring link.* For a four-bar linkage, the Grashof Condition is satisfied if $\boldsymbol{S+L \leq P+Q}$ where ***S*** is the shortest link, ***L*** is the longest, and ***P*** and ***Q*** are the other links. When the Grashof condition is satisfied, at least one link will be completely rotatable. Figure 1 show the sketch of the 4BL mechanism which can be described by nine independent

Figure 1. Four bar mechanism

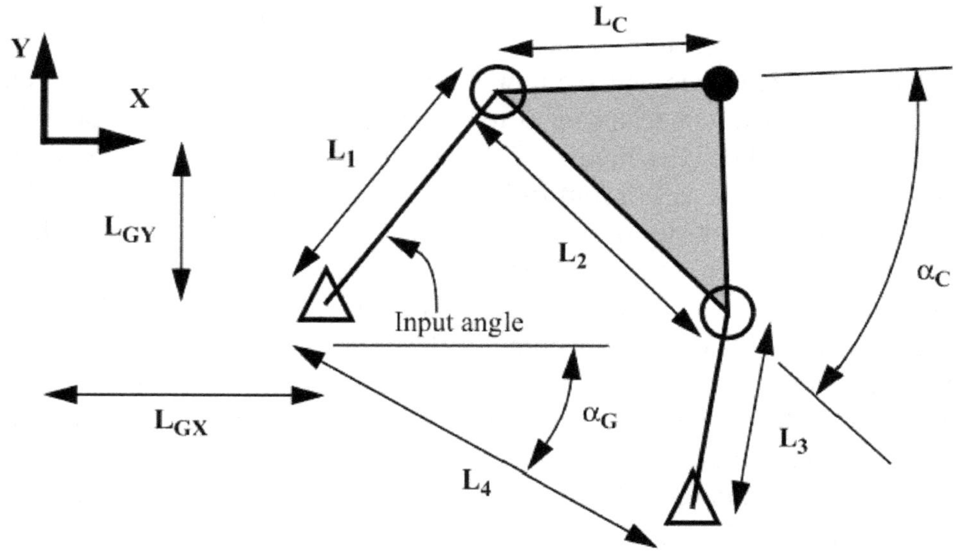

variables. The small triangles show revolute joints affixed to the ground, the open circles represent revolute joints that join two links, and the filled circle is the coupler point. Link 1 is the input link, the large shaded triangle is link 2, the coupler, link 3 is the output link, the follower and link 4 is the ground. Also shown are the offsets from the world origin and an angle that represents the angular offset of the coupler point.

Figures 2 and 3 show examples of the software found using YouTube. However these softwares were not interactive enough for the students to learn as it does not allow them to input numeric values and see the simulation (merely video based simulation).

METHODOLOGY

Computer based simulation tools has been employed by many higher learning instituions as an alternative to teaching engineering because it can be used to simulate and show realistic 2D to 3D models of a problem or process to simplify and explain the functioning of complex mechanisms. However a study conducted at the University of Rome stated that, "*although the integration between numerical computation and graphics leads to the generation of*

Figure 2. Four bar mechanism software sample 1

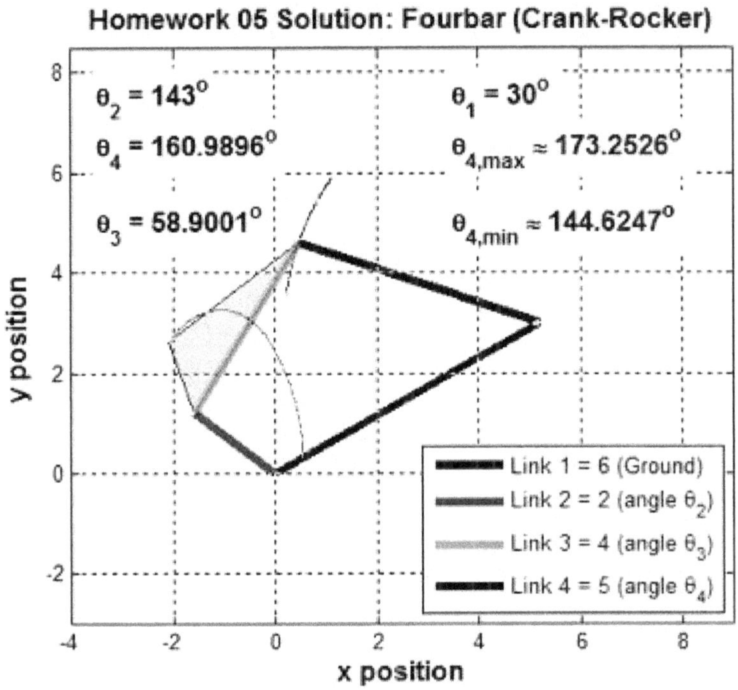

photo-realistic digital mock-ups, they are still far from the real context and the user has a limited interaction with them" (Paolo, Valentine, and Pezzuti, 2010). Furthermore this limitation can generate problems (non-conformities, unexpected behavior and appearance, etc.) when the designed products have to be integrated in the real world. To overcome this disadvantage, the development of new software tools, based on the mixing between the real world and virtual objects, seems to be the futuristic to computer aided learning. This section describes the design approach and methodology used in simulating the 4BL mechanism followed by an explanation of how it can help students to visualize and comprehend the problem while interacting with the simulation tool.

Most simulation of 4BL mechanisms that are available at present are designed using 2D graphics and may not be appropriate for visualization process due to the aforementioned reasons. In addition the workings solution of a given problem is not shown in details thus living the student to work on their own to understand. In our approach the student is able to interact with the 4BL mechanism in a real time 2D environment and is able to experiment

Figure 3. Four bar mechanism software sample 2

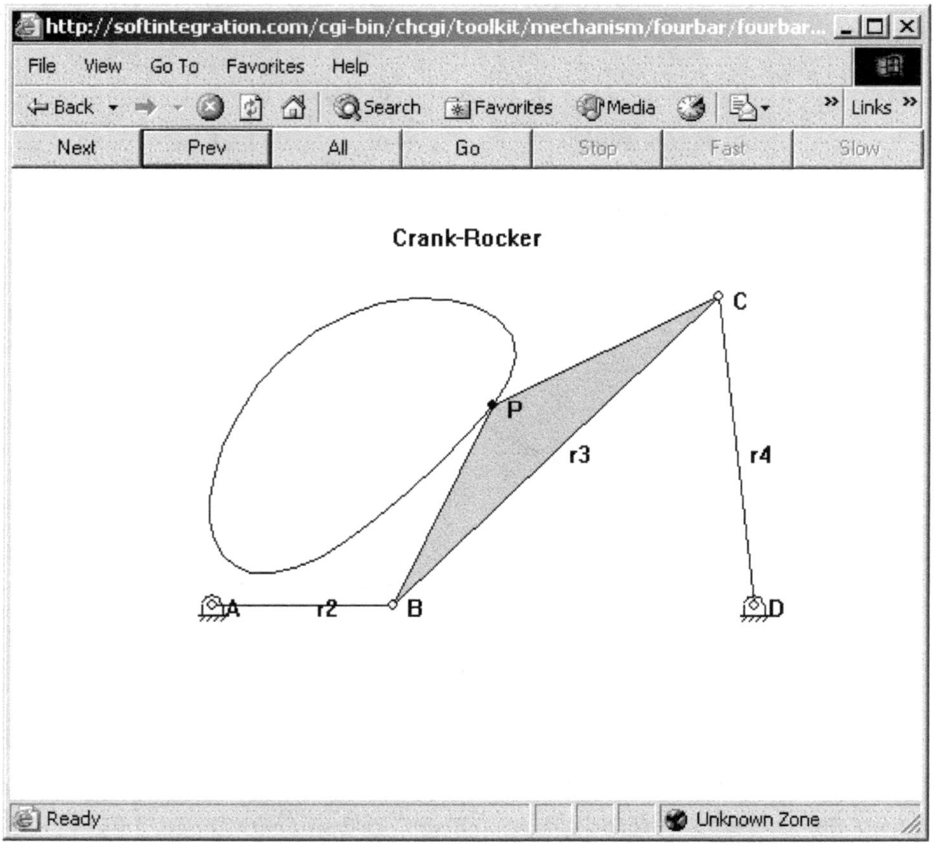

the parameter updates of the animated 4BL by changing different values that meets grashofs law. As for the development of the 2D model, we used MATLAB with graphical user interface as the main platform. This simulation graphical tool was modified from the original design by John Hawkins (some features were removed and new ones added). Figure 4 shows the main interface of the 4BL simulation tool that was designed and enhanced in this research.

The simulation is a graphical tool to visualize 4BL mechanisms and the instant centers that are produced by the mechanisms, and plots the path/trail of the instant centers. The paths of the instant centers are shown in Fig. 5 (blue lines) for the mechanism and the inversion of the mechanism (switching the coupler link and the ground). The paths represent two curves, that when one roles on the other, produces the same motion as the original 4-bar coupler that made them, e.g. the 4-bar can be synthesized by a 2-bar mechanism consisting of only the fixed and moving path. The moving path curve will have two points on it that travel in circles, just as the original coupler did on the 4-bar.

The user can select from eight different types of 4-bars, e.g. cranks and rockers or combinations of the two. Additionally the user can specify any link length of interest, numerically with slider bars, and do so while the animation is running (by pressing the "Reset lengths" button) shown in Fig. 5. The "Run" button allows the user to see the simulation after manipulating the parameters of the input angle and link lengths. The user may also save the animated image as a "jpeg" file at any time desired by clicking on the "Save as image" button. This could be useful for future reference.

In terms of graphical user interface and the simplicity of data manipulation we adopted the slider bar button. The slider bar button could be used to adjust the animation speed and input angle specification. When the user clicks the "Pause" button, the input angle may be specified by dragging the slider to move the mechanism. Appropriate colours were used to draw the students attention. For example the two small green triangles (Figure 5) show revolute joints affixed to the ground, the open circles represent revolute joints that join two links, instant paths are shown in "blue colored curves" and the red lines are the linkages.

In order to gauge the level of effectiveness of the simulation tool, the user has the opportunity to apply the knowledge gained through the lesson by simply going through the prepared exercises as typically shown in Figures 6 - 8. Three types of questionnaires are available namely multiple choices, true/false and fill in the blanks. The exercises are design in such a way that the students can interact with the system and keep on trying until they obtain the correct answer to all the wrong ones only.

On completion of the design, fifty-five students participated in the one week trial of the prototype of the TAPS MATLAB simulation tool, in August 2011. This followed an informal evaluation of an earlier, traditional based teaching by the subject matter instructor. As part of the one week trial of the TAPS MATLAB simulation, students completed a pre and post-test questionnaire. The questionnaires consisted of three sections:

1. Interest in, and knowledge of, aspects of the 4BL mechanism;
2. Access and experience with computers; and
3. Views of traditional teaching and learning.

Figure 4. The main interface of the 4BL simulation tool

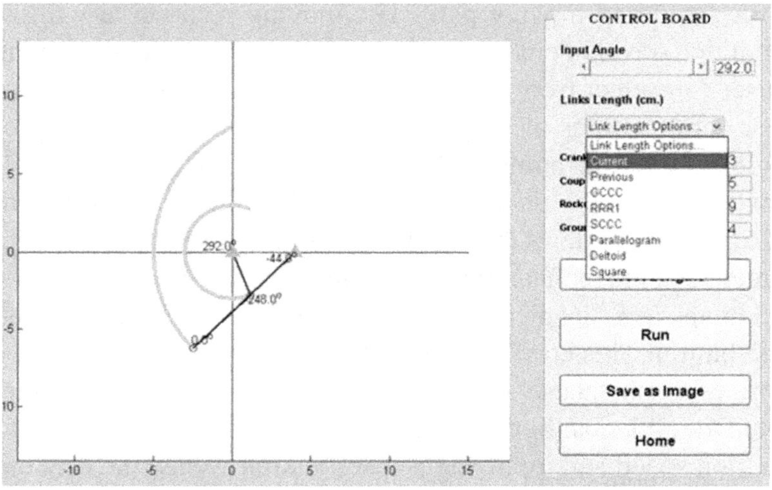

Figure 5. The paths of the instant centers

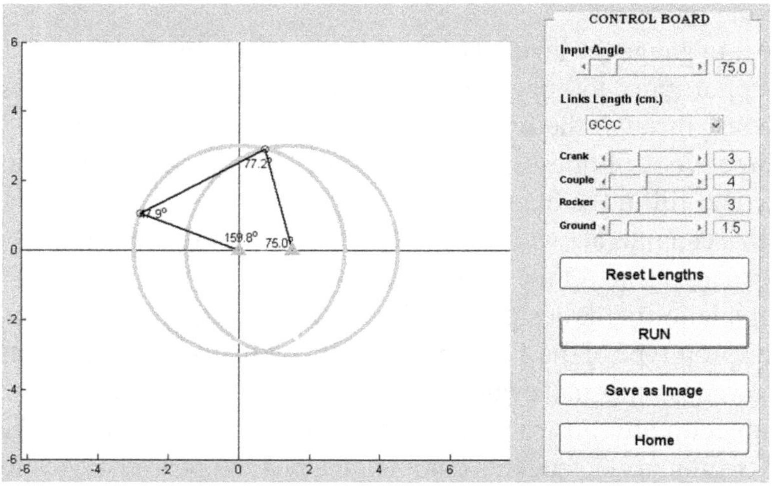

Figure 6. Multiple choice questions

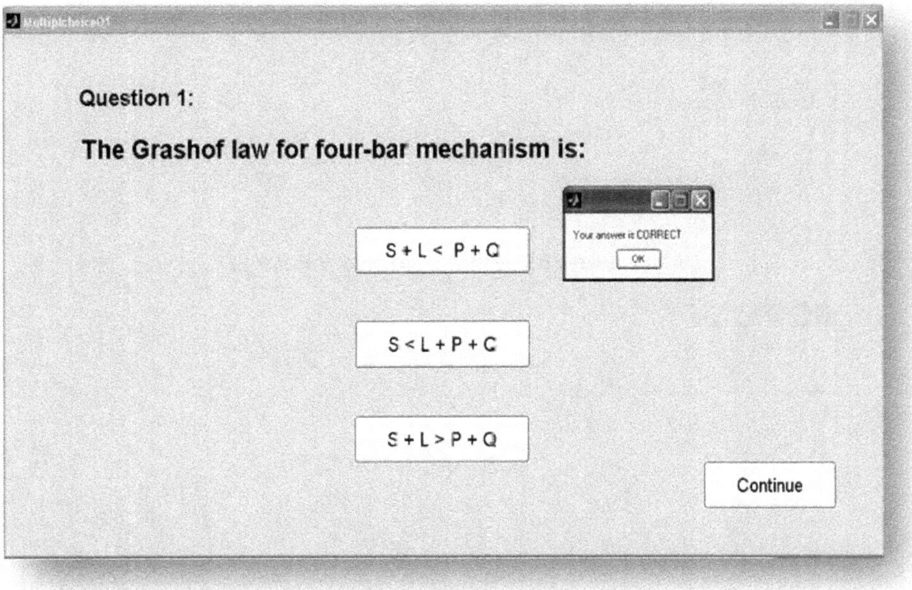

Figure 7. True and False questions

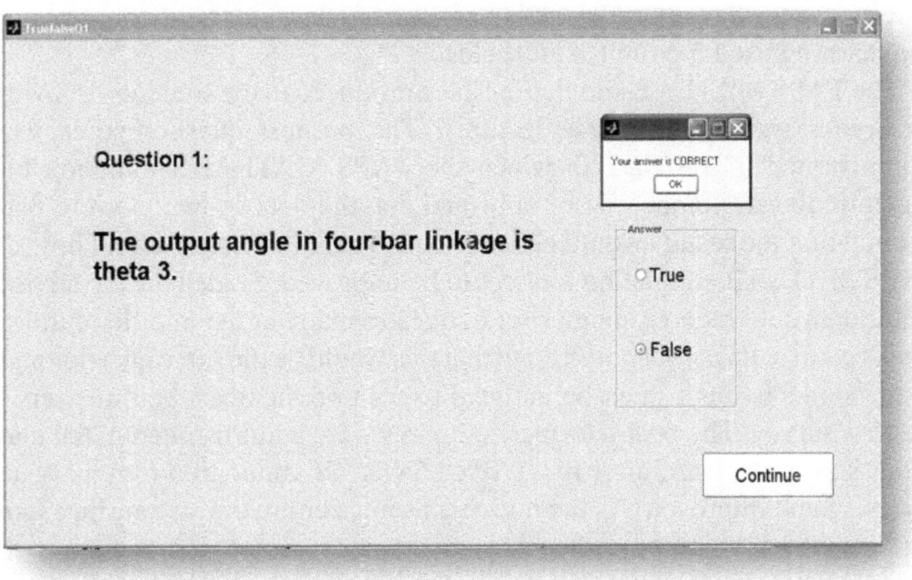

Figure 8. Fill in the blanks questions

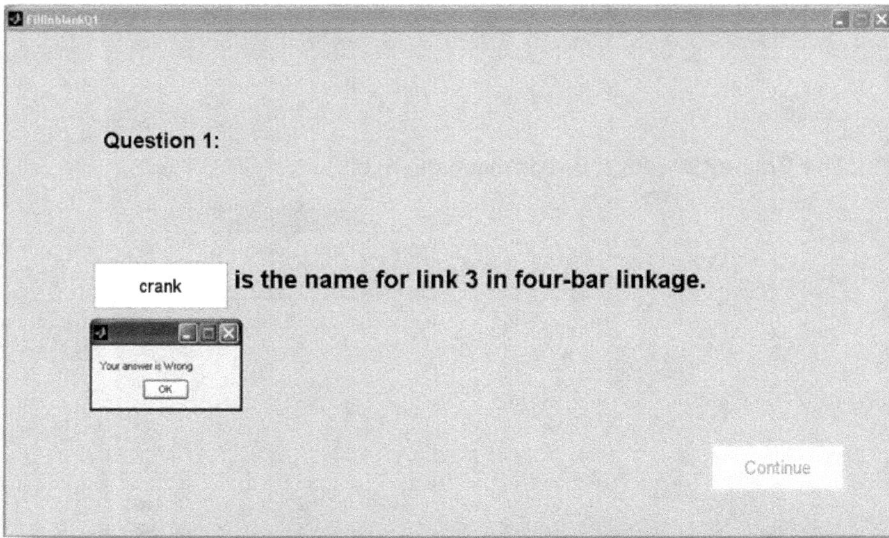

Answers to questionnaires in section (a) required a combination of ratings on 5 point Likert scales (1 strongly disagree to 5 strongly agree) and open ended questions. Section (b) contained some yes/no questions (for example, do you have access to computer at home?) as well as questions requiring an answer on 4 or 5 point scales. All questions on views of traditional teaching and learning used 5 point Likert scales.

The TAPS MATLAB simulation was introduced to the students following six weeks traditional course lectures. The pre-test questionnaires were administrated in the class in which the TAPS MATLAB simulation tool was introduced. Students were informed that the questionnaires were not a test and that the results would be used to provide information about how the TAPS MATLAB simulation tool could be improved. Students were advised that they would receive another set of questionnaire at the middle of the six weeks course. It was not pointed out that this would be the same questionnaire as the students might then be inclined to try to remember their answers to the first survey. The post-test questionnaires were administrated in the class about six weeks later, after the TAPS MATLAB simulation tool had been designed and improved. As the mid-term semester break was nearing, some students had gone home earlier thus attendance was not high enough in the initial class. Comparisons between pre and post-test responses were made using t-tests for related groups.

A focus group of 11 students was also conducted by an academic member. During this time the focus group was asked their overall impression and understanding of the TAPS MATLAB simulation tool and their suggestions for improving the tool.

RESULTS AND DISCUSSION

The majority of the students (80%) had access to a personal computer at home, rated their computer skills as average in using computer based learning software in their learning (59%), and had used at least one application at the foundation degree level course.

Pre and post questionnaire were available for 37 students (67%) as 18 students failed to complete a questionnaire at the end of the pilot study. Some students who failed to complete the questionnaire were added in the focus group. A comparison between the pre-questionnaire responses by those who completed both questionnaires and those students who only completed the first questionnaire revealed no significant differences in the responses of the two groups.

Comparisons of students' pre and post-test survey responses showed that after the students who had used the TAPS MATLAB simulation tool rated their knowledge ($t=-4.46$, $df=35$, $p<0.001$) (see Table 1). Students also rated their interest in using the TAPS MATLAB simulation tool as significantly greater after understanding the concepts of 4BL mechanism ($t=-2.27$, $df=36$, $p<0.05$).

To determine whether the TAPS MATLAB simulation tool aided the students to visualize the 4BL mechanisms, pre and post-test questionnaires asked students to solve selected problems of the subject matter. In the pre-test questionnaire average number of score was 2.6 with 25% of students able to answer at least two questions correctly out of four questions. This compared to a mean of 3.2 in the post test questionnaire with only 7% of students unable to answer two questions correctly.

The evaluation results presented here represents the preliminary phase of the assessment, examining the first group of students who participated. The results are motivating. However further formative evaluation results are required to see whether the results hold and are sustained across time and over different settings and with different student group is planned. On the basis of

Table 1. Mean and standard deviations of responses to the pre and post-test questionnaires

Question	Mean Pre Survey (Standard deviation)		Mean Post Survey (Standard deviation)	
How would you rate your present knowledge on the 4BL mechanism?	3.1	(.54)	3.4**	(.64)
How would you rate your present knowledge on the way the 4BL mechanism topic as learnt from the textbook?	3.1	(.80)	3.5**	(.69)
How would you rate your present knowledge on the way the 4BL mechanism topic as taught by the instructor?	2.9	(.70)	3.4**	(.72)
Scale used 1 = very poor; 2 = poor; 3 = average; 4 = good; 5 = very good				
Learning the 4BL topic is easy?	3.7	(.62)	3.8*	(.46)
In general I am interested in using the MATLAB CAS simulation tool?	3.0	(.10)	2.8	(.76)
I found the MATLAB CAS simulation tool to be useful and easy to use?	3.4	(.93)	3.2	(.96)
The MATLAB CAS simulation tool helped me in visualizing the 4BL mechanism better?	2.6	(.87)	2.4*	(.73)
I solve similar problems better after using the MATLAB CAS simulation tool?	2.5	(.10)	2.6	(.87)
Scale used 1 = strongly disagree; 2 = disagree; 3 = neutral; 4 = agree;5 = strongly agree				
Items where there were significant difference between pre and post-test questionnaires *p<0.05, **p<0.01				

these results it is suggested that the TAPS MATLAB simulation tool helps students in the visualization process. Additionally it allows students to interact with the TAPS MATLAB simulation tool to put knowledge gain in class into practice. The achievement of these learning outcomes was supported by the results from the evaluation, which showed that at the completion of using the TAPS MATLAB simulation tool students rated and demonstrated, that their knowledge about 4BL mechanisms and applying the right equations in solving the problem has significantly increased.

SUMMARY

The employment of computer software tools for modelling and simulating the operation and functions of mechanical engineering mechanisms is an essential pedagogic complement to modern teaching and learning. The study undertaken in this research is an important contribution in the educational domain of information technology and students and instructors of mechanical engineering department of UNITEN. Previous experience of instructors teaching the 4BL mechanism problem at UNITEN found that students had difficulty in understanding the basic concepts and applying the right equations. To solve this problem, the researchers propose a pedagogic approach based on the use of the TAPS MATLAB simulation tool to help students to comprehend and visualize the motions of the 4BL problem in an interactive manner. The evaluations results of the TAPS MATLAB simulation tool revealed that the tool was useful and could help the students in understanding the topic better.

The design of the graphical user interface using MATLAB was a great help to students in interacting with the TAPS MATLAB simulation tool as the students were familiar with such user interfaces due to their experience of surfing the Internet for related information. The use of appropriate colors and labels in the tool also helped students to grasp the concepts of 4BL mechanisms faster as opposed to the traditional method of learning. The inclusion of the three types of questionnaires (multiple choices, true/false and fill in the blanks) in the TAPS MATLAB simulation tool to gauge the students understanding was also helpful as similar existing tools available on the Internet do not provide such features.

ACKNOWLEDGMENT

The author would like to acknowledge the considerable contributions of Kather Ali a Masters in IT student who modified and simulation the MATLAB simulation tool.

REFERENCES

Huber, G., & Dietmajer, P. (2010). Four Bar: An educational synthesis and design tool for four bar mechanisms. *Proceedings of the 10th World Congress on the Theory Machines and Mechanisms*, 20–24.

Kihonge, J., Judy, M., & Vance, M. (2002). Spatial Mechanism Design in Virtual Reality with Networking. *Journal of Mechanical Design, 124*(3), 435–440. doi:10.1115/1.1481363

Manjit, S. (2007). *Development and Applications of Interactive Multimedia TAPS (Technology Assisted Problem Solving) Packages for Engineering.* PhD Thesis.

Manjit, S. (2008). Comparative Study of TAPS Packages for Engineering. *International Journal of Instruction Delivery Systems, 21*(3), 38 – 42.

Manjit, S. (2009). Visualizing Curvilinear Cylindrical Engineering Members in 3-D Environment with Motion Path. In *NCM2009: 5th International Joint Conference on INC, IMS and IDC*. IEEE.

Manjit, S. (2011). A 3-D Graphical Model for a Four Bar Linkage Mechanism Problem. In *First IRAST International Conference on Data Engineering and Internet Technology (DEIT)*. IEEE, SGU & IRAST.

Manjit, S., Ramesh, S., & Selvanathan, N. (2003). The Role Of Courseware In Engineering. In *32nd International Conference on Computers and Industrial Engineering*. University of Limerick.

Oleg, V. (2000). *Fundamentals of Kinematics and Dynamics of Machines and Mechanisms.* CRC Press. DOI: doi:10.1201/9781420042337.ch1

Paolo, P., Valentine, & Pezzuti. (2010). Interactive Multibody Simulation. *Journal of Theoretical and Applied Mechanics, 48*(3), 733–750.

Wikipedia. (2011). Retrieved from http://en.wikipedia.org/wiki/Four-bar_linkage

KEY TERMS AND DEFINITIONS

4BL: Series of rigid links with joins to form a closed chain.

MATLAB: MATLAB (matrix laboratory) is a multi-paradigm numerical computing environment and proprietary programming language developed by MathWork.

Simulation: Is the imitation of the operation of a real-world process or system over time: an object oriented model represents reality.

Visualization: Is any technique for creating images, diagrams, or animations to communicate a message. Visualization through visual imagery has been an effective way to communicate both abstract and concrete ideas since the dawn of humanity.

Chapter 5
Spur Gear Design and Analysis ATAPS Package

ABSTRACT

The subject of spur gear design problem in mechanical engineering (ME) deserves significant attention since the subject relies heavily on the combination of complex mathematics formulas, graphs, and tables. Many students are facing difficulty in understanding the subject of spur gear design in Universiti Tenaga Nasional (UNITEN). Hence, a study was conducted to enhance the learning experience of UNITEN's ME students in the subject matter. In this research, the deterministic, divide, and conquer algorithms were employed to improve the problem-solving technique. The combination of these algorithms provided room for students to make assumption based on their input rather than having the computer to do all the calculation and as well as to break down the complicated problems into sub problems. Additionally, it was designed to examine the effectiveness of using computer-aided software to improve the learning experience and comparing it to the conventional approach.

INTRODUCTION

The data collection procedures involved using both methods of teaching, which is the Computer-Aided Learning (CAL) approach and conventional approach. Both groups of students were assessed based on their prior computer experience, preferred learning styles (as a result of Felder-Solomon's ILS questionnaire), learning experience and comparing the result in solving spur

DOI: 10.4018/978-1-7998-0465-9.ch005

gear design problem with and without the CAL software. An analysis of the participants' results revealed that CAL software has helped ME students in solving spur gear design problem with a better approach. The first evaluation finds that more students from the group with Spur Gear Analysis Software were able to create a more cost and weight effective spur gear design problem as compared with the group without Spur Gear Analysis Software. The second evaluation finds that visual representation such as graph helps ME students in visualizing the concept and foreseeing the relationship between each parameter involved, making their learning experience easier and better.

The rapid growth of technology has made our life easier in one way or another such as the implementation of technology in education which is also known as Computer Aided Learning (CAL), Computer-Based Learning (CBL), and Computer Assisted Learning. The introduction of CAL has improved the conventional teaching and learning environment as CAL allows information or knowledge to be presented via multimedia such as image, text, audio and video (Sidhu et al., 2002). A study by Felder and Silverman (1988), has mentioned that the information presented in such way could cater for a diverse learning style, so students could benefit the most out of it. Hence, CAL is not something new and has also been implemented into many education fields which require more than just words to explain and present the theory or concept such as engineering education. In general, CAL has been used widely in the field of engineering as it involves heavily on the combination of mathematical formulas, free body diagrams and the effect of forces, where CAL is quoted by researchers as a solution to create a better problem-solving environment for engineering students. Hence, this work highlights recent work done in CAL and engineering education but specifically on Spur Gear Design in Mechanical Engineering. Additionally this work was also carried out to find research possibilities that could refine CAL to not only solve problem but also to improve the engineering student's learning experience (i.e. making assumptions/predictions).

BACKGROUND

Since the introduction of the Outcome Based Education (OBE) into the engineering programs by the Board of Engineers Malaysia (BEM), all engineering programs in Malaysia such as civil engineering, electrical and electronic engineering and as well as Mechanical Engineering program have to be OBE compliance in order to obtain full accreditation from BEM. Hence, Bloom's Taxonomy was developed in 1956 by Benjamin Bloom to measure and categorize the Outcome-Based Education (Eisner, 2000). Researchers has also found various problems in the teaching and learning process of engineering subjects which are further discussed in the section titled "Problems in Teaching and Learning Engineering Subjects."

With the emergence of problems in teaching and learning, studies by Sidhu (2010), Neumann, Neumann and Hood (2011) and Smetana and Bell (2012) suggested that CAL can solve the engineering problems and foresee it as the solution for future. This is because CAL has the ability to address different learning styles with its information presented in multimedia, 2-D and 3-D mechanic models environment, creating an interactive learning environment in teaching engineering subjects which require real-time visuals to demonstrate the concepts. Works done by Huang et al. (2011), Raveendran and Batmanabane (2012), Sidhu and Ali (2013), Vattam et al. (2014) and Sanchez at al. (2014) who incorporated CAL into the engineering learning environment have provided evidence demonstrating the fact that CAL outperforms the conventional learning method in term of student's learning outcomes. However, the existing CAL solution still does not promote student's learning outcome to a higher level thinking which is level 4, 5 and 6 (based on the Bloom's Taxonomy model). The findings gathered from the above authors have raised a research opportunity to create a comprehensive problem solving environment, so that engineers are equipped with a higher thinking skill upon graduation. There have not been much studies conducted in the Malaysian context in using CAL to enhance the understanding of engineering students in a difficult and challenging mechanical engineering subject which is the spur gear design. Therefore, it has led to this research, study of CAL to create a comprehensive learning environment to promote higher thinking skill in the mentioned perspective.

PROBLEMS IN TEACHING AND LEARNING ENGINEERING SUBJECTS

It is known that engineering subjects involve heavily in free body diagrams, theories, effect of forces which are comparatively hard to be explained with just words or lengthy explanation on the scenarios. In general various studies

Table 1. Reported problems encountered in teaching and learning engineering subjects

Authors	Study	Problems Encountered
Litzinger et al. (2011)	Engineering education should encompass learning experiences that allow students to construct deep conceptual knowledge apply key technical and professional skills fluently and engage in authentic engineering projects.	Students often fail to develop deep conceptual understanding due to the nature of the conventional learning experiences such as solving traditional, textbook types of problems With a large amount of content in engineering curriculum, there is time constraint in implementing and creating a learning environment for deep learning approaches to cover the entire curriculum Creating special facilities for effective learning experience requires financial resources
Direito et al. (2012)	Examines how undergraduates rate their proficiency in soft skills and the important of it for future employment.	Difficulties in relating and applying the knowledge to other domain. For example, using the knowledge learnt in class for industrial work.
Stieger et al. (2014)	The challenges in teaching in front of large audiences and way to improve engineering education in shifting from teacher-centered to student-centered education.	As the number of students increase and at the same time shrinking funds, lecturers are confronted with large audiences (>700 students), there are difficulties in conveying the subject matter of mass to individual student and be understandable for everyone

have revealed the same issues or problems faced in the engineering education. Some general problems faced by lecturers and students in teaching and learning engineering subjects can be summarized as following in Table 1. From the literature review, the problems encountered by lecturers and students can be streamlined into three categories which is the ability to spatial visualize, ability to make assumption and ability to understand as categorized Table 2.

From the aforementioned statements (Table 2), it is clear that gaining complete understanding in the subject matter is what it takes for engineering students to deal with uncertainty and solve complex problems. Only students that are equipped with good conceptual understanding will be able to explain their solution rather than just manipulating the formulas to solve the problems (Yadav and Barry, 2009).

APPROACHES TO IMPROVE ENGINEERING EDUCATION

With a variety of learning styles displayed by engineering students, many different learning approaches have been developed and designed to cater

Table 2. Categorization of problem encountered

Ability to Spatial Visualize
Students do not see the relevance of the learning materials when it is being taught out of context (Mickleborough and Wareham, 1993, cited in Ditcher, 2001, p.25) Inabilities to spatial visualize the relationship between reality and theoretical concept (Crown, 2001)
Ability to Understand
Difficulties to visualize some of the more difficult concepts in engineering to apprehend theory and practical (Sidhu et al., 2002) Students are unable to comprehend how the knowledge is structured (Ellis and Turner, 2003) Difficulties in solving problem when it is asked differently than the ones in their exercises and instructions (Ellis et al., 2004) Students often fail to develop deep conceptual understanding due to the nature of the conventional learning experiences such as solving traditional, textbook types of problems (Litzinger et al., 2011)
Ability to Make Assumption
Students are lacking of estimation skill, which is vital in engineering education (Linder, 1999) Even students in the senior year require help with estimation to solve the problem effectively (Dym et al., 2005) Engineering students are inadequate in making estimation even on the simplest estimation problem (Shakerin, 2006)

for a wide range of students. However, problems in engineering student's learning environment still persist such as analyzing and making assumption in solving engineering related problems. Over the years, engineering education has evolved from the conventional paper and pen approach in teaching to computer based learning, games-based learning and many more. Although all these approaches are designed to improve the engineering education but not much has improved on the state of affair.

Additionally most of the approaches mentioned above, highlights on the active engagement of students in the learning environment. This can be found in Active Learning where it engages students in classroom activity like reading, discussion and writing (Bonwell and Eison, 1991; Ragains, 1995; Keyser, 2000 & Prince 2004). Besides, Cooperative Learning also involves the student's engagement in term of working in teams to accomplish a task (Johnson, Johnson and Smith, 1991; Felder and Brent, 1994 & Prince, 2004). Computer Aided Learning (CAL) and Game-Based Learning have the student engagement element too, where the use of computer or video games allows interaction and high degree of student control, making it engaging and interactive (Perrotta et al., 2013). However the approach that has been applied extensively to engineering namely CAL which is discussed further in the next sub section.

Figure 1. Different learning terminology

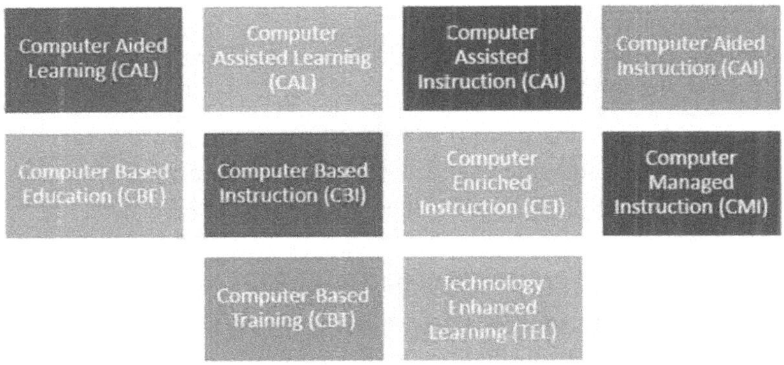

COMPUTER AIDED LEARNING

With the growth of technology, Computer Aided Learning (CAL) has become a pervasive approach in problem-solving. CAL is also known by different names and abbreviations like Computer-Based Instruction (CBI), Computer-Based Learning (CBL) and Computer Assisted Instruction (CAI). Figure 1 shows different terms and abbreviations commonly used for computer-aided learning.

However all of the above mentioned learning terminologies have more or less the same meaning which is the involvement of technology in the teaching and learning process, by means of integrating computer software into the learning environment. CAL is defined as the learning procedures and environments guided by computers but the main idea of understanding CAL is the interaction, whereas interaction limited between the user and the content (Schittek et al., 2001). According to Daunt et al. (2013), any use of computer technology to improve one's knowledge is also known as CAL. Integrating computer in the teaching and learning activities is known as CBI (Serin, 2011). Overall, CAL is a term used to describe the replacement of a human instructor in the education or training of a learner/student with the computer software application (Sidhu, 2010). In general a number of applications have been designed to teach engineering using learning terminologies stated in Figure 1.

CAL IN SPUR GEAR DESIGN

Spur Gear Design is one of the most challenging topic in the Machinery Design subject as it involves heavily on mathematical formulas, tables and graphs that are interrelated among each other. In analyzing the stress distribution of the spur gear design, only one formula is used to compute the bending stress of the machinery. Although there are more than 10 variables that are inter-related in the formula and in most problems, only 4 to 5 values will be given in the question and students have to make estimations for the rest to solve the ill-defined problem. Students are required to have a thorough understanding of the subject beforehand, especially the relationship between each parameter involved in the formula used, so that they can make suitable assumptions like the materials used, "shockiness" of the machinery, and quality index of the spur gear set to identify the bending stress analysis of the spur gear. In a normal practice, students that are equipped with the visualization skill will have the advantage in computing and identifying the stress distribution easily, as they can relate the reality and the theoretical model in mind before putting the values into the formula for design. While those students that lack visualization skills will have difficulties in computing the stress distribution of the spur gear set.

Determining a cost effective spur gear design is vital in the design of power transmission applications such as cement mixer drum driver, printing press, punch press, steel mill drives and many more (Gologlu and Zeyveli, 2009). Therefore efforts have been concentrated on improving the problem-solving approach to design an effective gear design. Researchers have designed two advanced optimization algorithms known as particle swarm optimization (PSO) and simulated annealing (SA) to find the optimal combination of design parameter for a low weight spur gear train (Savsani, Rao and Vakhria, 2010). Another study by Padmanabhan et al. (2010) optimizes the spur gear pair design using Genetic Algorithm (GA) and analytical tool MITCalc, to maximize the power efficiency and minimize the overall weight. Many efforts have been focused on optimizing the spur gear design while little attention was given to improve the learning curve of mechanical engineering undergraduates in the subject matter. This work is aimed to motivate and foster more works to be done in spur gear design in term of education to meet the need of mechanical engineering undergraduates in spur gear design.

In the era of the 4th industrial revolution, numerical simulation tools would be very useful for engineers to analyze and predict the condition of real-world physical systems. Various computer packages have incorporated computer technologies, to solve mechanical engineering problem in Machinery Design either for industrial or education purposes. They have been designed to cater to the trend towards visual data analysis, but not many have been completely employed into the learning environment. A list of such packages and a comparative are presented in (Maisarah, 2016). However, most of it is designed for industrial purposes and it is not widely employed in the classroom teaching. According to Ho et al. (2002) the challenge for this includes the cost in developing CAL materials, which does not yield any financial return to companies that have invested a lot in producing these materials. The insufficient materials available in free CAL packages for machinery design made it hard to be used for teaching due to lack of functionalities. As for the commercial computer software, it creates confusion and complication among students with the commercialized functions and materials provided in the software. This will eventually discourage students in learning the subject matter. Consequently it is permissible to a certain extent to produce a tutoring and problem-solving package that can aid students in a more effective manner than the existing CAL packages for spur gear analysis in machinery design.

LEARNING STYLE

The purpose of this section is to describe and determine the major characteristic affecting student learning style in order to incorporate preferred learning style in the design and delivery of the CAL software. In a classroom environment, every student exhibit a different learning style but the classroom content will only be delivered in a specific type of teaching style. This scenario has affected the student's learning outcomes which the specific teaching style(s) has favored some students and the others are neglected. A description of learning style models and instruments available to measure one's learning style is explained in (Manjit, 2006). As for this research we selected the Felder and Silverman (1998) Index of Learning Styles (ILS) which consists of 44 items with 11 items for each style since it has been used and have a high validation rate and was found to be suitable for engineering education.

METHODOLOGY

Since this study has the aim to identify whether interactive software will improve the learning curve of engineering undergraduates specifically in the subject of spur gear design and to identify the effect of such intervention, quasi-experiment with non-equivalent-group experiment setup was adopted to compare and analyze the changes between both of the pre and post result.

Quasi-experiment is an empirical study to determine the causal impact of an intervention on its target population (Freemantle et al., 1998; Kampenes et al., 2008). Quasi-experiment some sort shares many similarities with traditional experiment design or better known as randomized controlled trial. However in quasi-experiment, the population is not randomly selected, the treatment conditions are all determined or assigned by the researcher. There are four proto-typical quasi-experimental designs which are the one-group 'pretest-posttest' design, the interrupted time-series design and nonequivalent-group design and the regression-discontinuity design.

Since this research is an intervention study of integrating computer software into the traditional teaching method which focus on pen and pencil method, the quasi-experiment nonequivalent-group design is used to measure the effect of such intervention. Nonequivalent-group design is an experimental design that compares across different participants and the effectiveness of the treatment is assessed by comparing the performances of the participants across the different groups (Reichardt, 2009).

Sample and Sample Size

The selected sample target is mechanical engineering (ME) students in UNITEN that enrolled in the Machinery Design class. This focus group is selected because studies have shown that among all of the engineering courses, ME instructors and students are facing the most problems in teaching and learning because it involves the combination of mathematics, free body diagrams and effect of forces which is rather hard to be explained and illustrated via the traditional teaching method, pen and pencil. ME undergraduates from Machinery Design are selected because according to lecturers in UNITEN, this 3rd year subject is able to not only assess the students understanding but also their skill in evaluating and creating during problems solving. Besides, this subject acts as the foundation for students in designing their own machinery or project during their final year, so it is important to ensure students have a thorough understanding of it before proceeding to their 4th year engineering program. Therefore the research sample was targeted on ME students in UNITEN that were learning about the spur gear design problem in their Machinery Design subject. As for the sample size, it was calculated based on the formula given and number of engineering student enrolled in UNITEN for mechanical engineering course which was obtained from the registrar. Since this study only focused on 3rd and 4th year mechanical engineering students, therefore 2 out of 4 from the total number of 1,421 of engineering students, which is 711 are put into consideration for this study. The total of 711 is 0.07% of the total number of enrolment in UNITEN.

Experimental Setup

Different experimental setups on the pre- and post-test will be explained in brief. This experiments took place in UNITEN specifically College of Engineering (COE).

Pre-Test

For the pre-test experiment which uses conventional pen and paper method, it took place during the Machinery Design subject's tutorial class which was handled by a tutor. The tutorial class is generally conducted in a fully air conditioned classroom and the environment is comfortable and exactly the same as regular classroom area. Having the pre-test conducted during tutorial class enabled students to take the test as normal quiz instead of being stressed out and pressurized by a new environment, which will eventually affect their performance in answering the questionnaire. Therefore it can be concluded that the environment was suitable for the pre-test which only involved the conventional pen and paper to take place without major disturbance or distraction by the environment.

Post-Test

As compared to the pre-test, post-test's setup is a little more tedious. Instead of conducting the test in classroom, post-test was conducted in the computer labs in COE, UNITEN. Two computer labs were used to accommodate the number of students, so every student had access to a computer each to answer the questionnaire individually. The Spur Gear Design Analysis Software was installed into the computers a day before the testing to ensure it has zero technical flaws in running the application on Windows XP Professional (Service Pack 3) operating system. A projector was used to show a brief walk-through of the application conducted by the facilitator before the students began solving the spur gear problem with the Spur Gear Design Analysis Software.

Data Collection

In this research, semi-structured interviews were conducted with the lecturers who have been teaching Machinery Design subject in COE. Interview questions, focusing on the topic of spur gear design; problems faced by lecturers and students in the topic and also proposed solution in tackling the issue. Interviews were recorded in written notes and then analyzed to identify the common themes and points of differences mentioned by the lecturers. Besides, covert observation were used to observe the way students react to the Spur Gear Design Analysis Software and using it to solve the given spur gear problem. This was to ensure that the 'live' data can be collected and recorded in written notes from the students in respond to the software rather than relying on second-hand accounts.

As for the quantitative data collection, there are two main phases with two different questionnaires. Phase 1 is the pre-evaluation phase where students' information was gathered while Phase 2 covers the testing phase on control and experimental group with the Spur Gear Design Analysis Software. The 2 sets of questionnaires focused on i.e. 1 pre (Students' Background Questionnaire, Prior Computer Experience Questionnaire, Learning Styles Questionnaire using the Index of Learning Styles (ILS) from Felder-Silverman Learning Style Model and Learning Experience in Spur Gear Design Questionnaire) and 1 post (Feedback on Spur Gear Design Analysis Software Questionnaire). The pre questionnaire format is adopted and utilized to identify the potential participants for the study and also to weed out exceptions such as student who did not meet the required CGPA or passing the pre-requisite subjects.

The post questionnaire format is adopted and utilized to measure the effectiveness of Spur Gear Design Analysis Software to help them in solving the spur gear problems.

Spur Gear Design Analysis Software

Spur Gear Design Analysis Software is developed to include graph plotting tools to enable students to plot graphs according to the parameters involved, giving a visual representation of the relationship between parameters. The visual representation such as graph is known to be helpful to engineers to visualize the concept as most engineering learners favor visual learning method (Felder and Silverman, 1988). However, if users are having difficulties in solving the problem, the student can choose the tutorial function in solving the problem. It guides students to solve the problem in a step-by-step manner showing them how the answer is obtained. Besides, a calculator function is included to provide a complete interface with all the tables and graphs included to solve the spur gear design problem without the going through the step-by-step method. These efforts have focused to improve the problem solving approach of engineering students in spur gear design so as to support the acquisition of theoretical knowledge that will be applied to industry upon graduation.

The algorithm used in the Spur Gear Design Analysis Software is to provide a problem solving environment that easier and faster without the help of a human instructor. There are two algorithms adopted for this problem solving approach, which is the deterministic algorithm and divide and conquer algorithm which are explained in details in (Maisarah & Sidhu, 2013). Deterministic algorithm has been broadly employed to various fields such as communication (Malekpour et al., 2008), computer science (Chen et al., 2011; Matousek & Schwarzkopf, 1996), engineering (Chen et al, 2011; Errico et al., 2007) and neurocomputing (Angulo and Torras, 2002), because it is one of the most practical approach that is feasible in a computer. Having deterministic algorithm which produces the same output at all time, it would be helpful for engineering students in solving spur gear problems so that a consistent result such as a cost and weight effective spur gear design can be produced time after time.

According to the study by Smith (1985), the principle of divide and conquer algorithm can be as simple as breaking down a problem that cannot be solved directly into sub problems to be solved independently, then by combining the sub problem's solutions to compose the solution to the original problem. This algorithm attempts to make a complicated and difficult problem into smaller manageable problems, then solve the smaller problems individually, and lastly reassembling the original problem with the solutions from the simple problems. So it can be understood and solved using a simpler approach instead of spending longer hours to work out the complex problem by itself. The reason that divides and conquer algorithm is adopted in spur gear design problem solving because it has a structural simplicity which promotes understanding, computational efficiency, parallel implementation and the diversity of applications (Smith, 1985). It has also shown to have a faster and higher quality result based on the study by Stach, Kurgan and Pedrycz (2010) and Arbenz (1992).

By employing the combination of deterministic algorithm and divide and conquer to solve the spur gear design problem, it provides room for students to make assumption based on their input rather than having the computer to do all the calculation and as well as to break down the complicated problems into sub problems to be solved easily.

Graph Plotting Function

Spur Gear Design Analysis Software examined various interactions for a tutorial and problem-solving topic on spur gear design in the machinery design subject. The aim of this graph plotting function is to enhance the understanding of the relationship between parameters in spur gear design with the visual aid provided such as graphs.

Apparently the method of teaching spur gear design can be tedious, difficult, time consuming and requires instructors to repeat the topic several times before the student understands it. In addition, the relationship among all the parameters involved in spur gear design is hard to be portrayed and explained with merely words and sentences. Thus, in order to overcome this issue, graph plotting function was used to illustrate the relationship between two parameters or more (refer to Figure 2).

Since engineering students favor visual learning method, graphs are the easier and direct way in showing the relationship between the parameters. In the graph plotting interface, the design is a straightforward menu design where left-side shows the parameters selection to be used in plotting the graph, which will be presented in the right-side of the interface. At center bottom, there is also a table with the data used to plot the selected graph, providing another way for students to understand the relationship. The parameter selection uses the drop down menu to list down all the parameters involved in Spur Gear Design, which is a total of 31 parameters including the cost and weight parameter. Cost and weight parameters are included to enable student to relate

Figure 2. Graph plotting interface

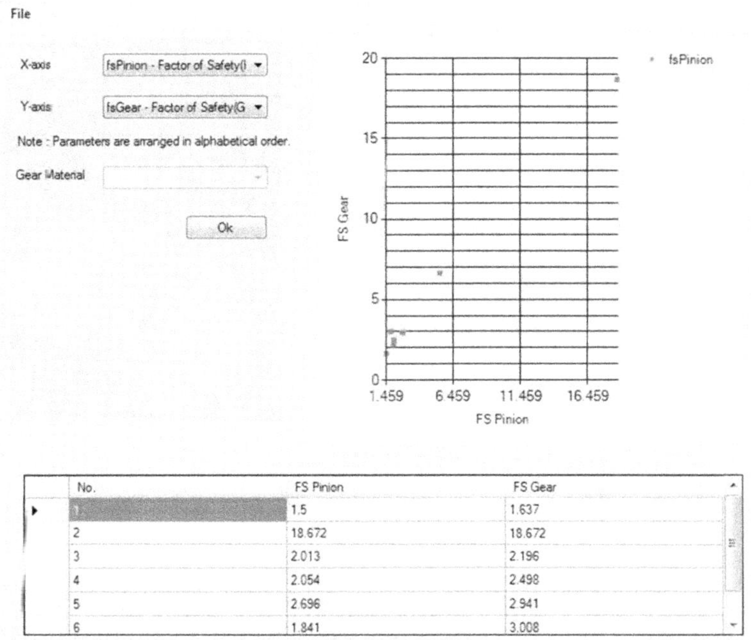

their spur gear design to the cost and weight factor, in creating a cost and weight effective design. Once the parameter selection has been confirmed, graph is plotted based on the parameter selection on X and Y-axis. Students may also approach the 3D graphical tutorial of the gears (refer to Figure 3) which is a straightforward interface with multiple windows arranged in a step-by-step order, prompting students to read information and visualize the particular gear in 3D with the gear's frictional part shown in red color.

DISCUSSION ON KEY FEATURES OF SPUR GEAR DESIGN ANALYSIS SOFTWARE

In general the Spur Gear Design Analysis software developed in this study was based on problem faced by students and lecturers in teaching and learning the spur gear design subject. For example, students have difficulties in visualizing the relationship between parameters involved in spur gear design; the graphs plotting features were developed since it is generally accepted as a simple and easy visual representation of a concept (Evaluation Research Team, 2008;

Figure 3. 3D graphical visualization of gears

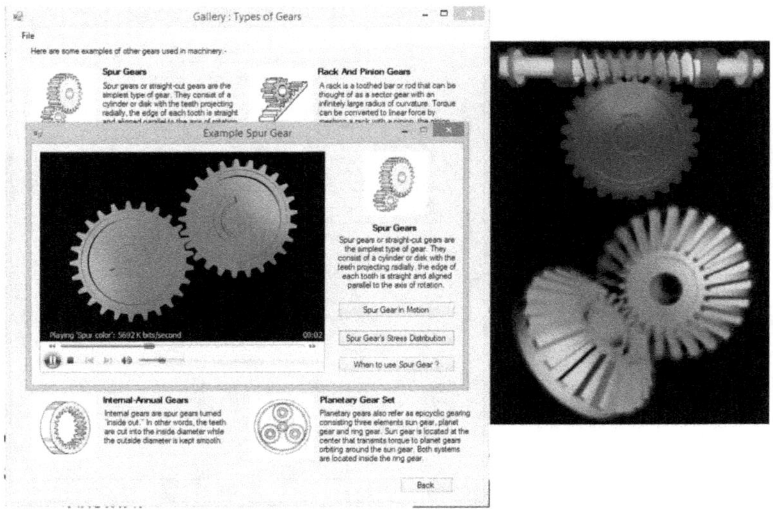

Arcavi, 1999). This also explains the relationship between the parameters involved in spur gear design. Besides, the graphs can be plotted immediately on the screen with any parameters, which is more convenient as compared to using the conventional paper and graph paper method in plotting graphs. According to Felder and Silverman (1988), most engineering students are sequential learner. Hence, the tutorial step-by-step guide function was included in the design of Spur Gear Design Analysis Software to guide students in solving the spur gear design problem in a systematic approach.

As suggested by Mickleborough and Wareham (1993, cited in Ditcher, 2001, p.25) that engineering students are having problem in associating the subject learnt in class with the real world engineering challenges. In spur gear design, other than the factor of safety of the gear set, cost and weight are also put into consideration in designing an effective solution, so it can be employed for industrial usage. Therefore, the cost and weight parameters were included in the Spur Gear Design Analysis Software whereby estimated cost and weight were shown in the solution. This enables the student to understand that the changes of certain criteria in spur gear design will also affect the cost and weight of it.

Results and Data Analysis: Statistical Tests

Statistical test is utmost important to analyze the data collection and determine the significant difference of the intervention in this study. There are multiple statistical tests used to determine the relationship between variables and the common ones are Chi-Square, Pearson Correlation and Regression. All of these statistical tests are widely used by researchers. In this research, Chi-square test is used due to the nature of data collected which is nominal and ordinal data and Statistical Package for the Social Sciences (SPSS) was used to conduct Chi-Square test. For Chi-Square, authors like Lan, Sung and Chang (2007), Huang, Huang and Hsieh (2008), Barnard et al. (2009) and Miranda (2014) have also applied Chi-Square in their study for data analysis in education.

Students Background Questionnaire

The feedback was obtained from 108 mechanical engineering (ME) students. The result indicated that 37.9% (41) of the respondents fall under the CGPA 3.01 to 3.50 and 36.4% (39) has a 2.51 to 3.00 CGPA. While, 25.7% of the respondents fall under the CGPA category of 3.51 and above, which is 17.9% (19), 5.7% (6) in CGPA of 2.01 to 2.50 and 2.1% (2) in CGPA 2.00 and below. Such feedback is used to segment the ME students into control and experimental group accordingly so that a fair number of students from each CGPA categories are segmented into the 2 groups.

Prior Computer Experience Questionnaire

Although the result shows about 98.1% of the respondents have access to computer while a minority of 1.9% does not have access to computer, but all of the respondents (100%) reported that they enjoy using computer and also use computer to study. Among the tasks the respondents performed with the computers are such as searching related information in internet, computing mathematical calculation, doing online discussion and others. It is clear that computer plays an important role in the students' learning environment as 96.3% of the respondents used computer to search related information in internet via computer, so that they will have better understanding of the subject matter and engineering concepts learnt in class. Besides, 61.1% of the respondents reported that computer was also used for online discussion after class among peers and tutors, while 46.3% used it to compute mathematical formula and 3.7% indicated that they use computer to access online notes, past year papers, research on the topic learnt in class and to complete their assignments.

Learning Styles Questionnaire using the Index of Learning Styles (ILS)

The results of the student's learning style based on ILS indicated that mechanical engineering students' learning style incline towards visual, active, sequential and sensing. Therefore, the Spur Gear Design Analysis Software is designed in line with the mechanical undergraduates learning preference so they can benefit the most out of it. To help visual learners, engineering concepts were represented with graphs in order for them to see the cause and effect of the concepts, whereas with the step-by-step tutorial feature enables sequential learners to understand the problem solving steps better. Both active and sensory learners benefit from hands on experiments in the software by experimenting with multiple values to observe the changes towards one another.

Learning Experience in Spur Gear Design Questionnaire

The Learning Experience in Spur Gear Design are divided into three parts which are about spur gear design, formula for spur gear design and the online material and software available for spur gear design. Result about the subject on spur gear design revealed that 50% of the respondents agree that the lectures and tutorial are easy to understand but they still prefer to have more examples given in the class which is a total of 88.9%. Besides, 90.7% respondents agree to have step-by-step explanation in textbook to solve the problem for safety factor in spur gear design. Hence this suggested a step-by-step tutorial function to be included in the Spur Gear Design Analysis Software to guide students in solving the problem in a more systematic approach. Although the respondents agree that the lectures and tutorial are easy to understand but 48.1% of the respondents are having difficulties in solving the spur gear problem based on what they have learnt.

The result about formulas for spur gear design showed that 35.2% of the respondents agreed that they have problem in visualizing the relationship between each parameter in the formula. While 51.8% have the tendency to mix up all the tables and graphs involved, as there are more than 10 tables and graphs involved with multiple options provided in designing a cost and weight effective spur gear design. Besides, with the numerous numbers of tables and graphs, 40.8% agreed that they have issues in making assumptions accurately based on the problem given. Result about software and online material for spur gear design revealed that 48.1% of the respondents agree that there is related software in the internet for Spur Gear Design Analysis but most of the software are not free, as reported by 57.4% of the respondents. Additionally 77.8% of the respondents also agreed that they need computer software to help them in visualizing the relationship between each parameter and 59.3% of them enjoy using computer aided software to learn. Hence, the results indicate that software for spur gear design is helpful in learning the subject matter. In terms results of learning experience with Spur Gear Design Analysis Software, 87.3% agreed that the software is a useful tutorial aid in learning the subject matter and 87.3% thinks that the software can be incorporated into the course such as to be used in the lab. Besides, 90.2% agreed that software can be used for revision purposes at home too as the material in the software is similar to the subject content in class. After using the software, 50.7% of the respondents find solving the spur gear problems are easier with the software than manual paper and pen calculation. Lastly, 84.4% of the respondents would like to use the software again in solving spur gear problems.

Result Without Spur Gear Design Analysis Software

Figure 4 shows the result for control group testing (without Spur Gear Design Analysis Software or better known as the conventional paper and pen method) in solving the given spur gear problem, has showed that 51.85% (28) of the mechanical engineering students were able to solve the question with a correct answer while 48.15% (26) provided an incorrect answer.

Results of the Control Group With Spur Gear Design Software

As for the experimental group (see Figure 4), (with Spur Gear Design Analysis Software), it is clearly shown that a higher number of mechanical engineering students were able to answer the spur gear design problem correctly, which is a total of 70.37% (38), whereas 29.63% (16) of the respondents were unable to provide a correct answer.

Figure 4. a) Result without spur gear design analysis software; b) result with spur gear design analysis software

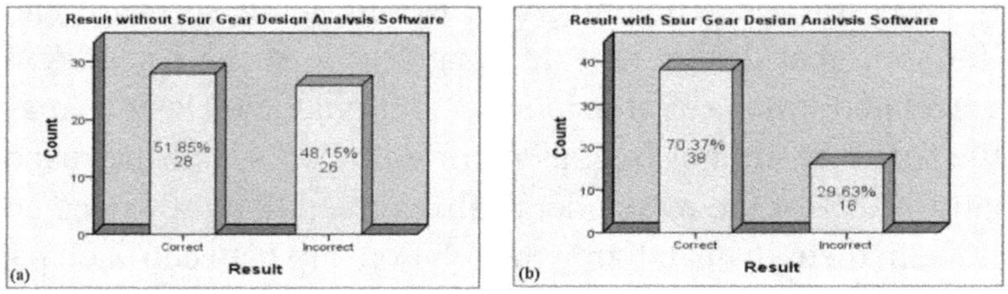

Results of the Comparison With and Without Spur Gear Design Analysis Software

Comparing the result in answering the spur gear design problem with and without the Spur Gear Design Analysis Software is tabulated in the bar graph above (Figure 5). It is clear that there is an increment in the percentage of correct answer from 25.93% (28) to 35.19% (38) in solving the spur gear design problem with the Spur Gear Design Analysis Software. Besides, the percentage of incorrect answer has dropped from 14.81% (16) to 24.07% (26).

A chi-square test with $\alpha = .05$ level of significance was run to determine the relationship between the usage of computer-based software and results in solving spur gear design problem. There was a strong, positive correlation between the two variables, which was statistically significant, $(1, N=108) = 3.896, p=.048$. A larger proportion of 57.6% reported that there is improvement in answering the spur gear design problem with the software compared with only 38.1% with conventional method. Computer aided learning has shown significantly different in understanding and solving the spur gear design problem as compared to the conventional paper and pen method.

SUMMARY

In general, computer software such as Spur Gear Design Analysis Software in this study is useful in learning challenging mechanical engineering subject as it allows the concept to be demonstrated in other format such as graph instead of just lengthy and wordy explanation. Besides, it can be designed in a sequential problem solving approach to match the learning preference of mechanical engineering students as found from ILS, making the students benefit the most

Figure 5. Comparison results with and without spur gear design analysis software

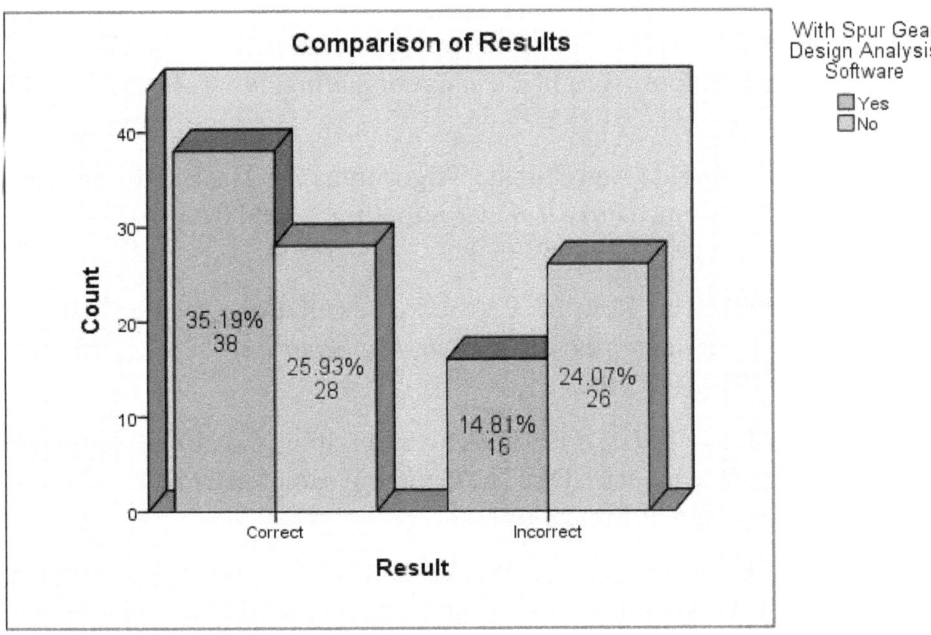

from the learning process. This work presented an approach in learning and understanding the topic of spur gear design amongst mechanical engineering students. Although numerous software can be found in the literature in gear design but very few is designed for education purposes. The finding from this study may create opportunities for researchers to explore the challenging topic of spur gear design to improve the learning experience of mechanical engineering students instead of the approach only in optimizing the gear set.

ACKNOWLEDGMENT

The author would like to acknowledge the significant contributions of Nur Maisarah Bte Shahril Khuzairi a Masters in IT student who designed and developed the Spur Gear Design and Analysis ATAPS package.

REFERENCES

Angulo, V. R., & Torras, C. (2002). A Deterministic Algorithm That Emulates Learning with Random Weights. *Neurocomputing*, *48*(1-4), 975–1002. doi:10.1016/S0925-2312(01)00695-6

Arbenz, P. (1992). Divide And Conquer Algorithms For The Bandsymmetric Eigenvalue Problem. *Parallel Computing*, *18*(10), 1105–1128. doi:10.1016/0167-8191(92)90059-G

Arcavi, A. (1999). The Role of Visual Representations in the Learning of Mathematics. *Educational Studies in Mathematics*, *52*(3), 215–241. doi:10.1023/A:1024312321077

Bonwell, C. C., & Eison, J. A. (1991). Active Learning: Creating Excitement in the Classroom. Washington, DE: The George Washinton University (ERIC Clearinghouse on Higher Education).

Chen, J., Feng, Q., Liu, Y., Lu, S., & Wang, J. (2011). Improved Deterministic Algorithms For Weighted Matching and Packing Problems. *Theoretical Computer Science*, *412*(23), 2503–2512. doi:10.1016/j.tcs.2010.10.042

Crown, S. W. (2001). Improving Visualization Skills of Engineering Graphics Students Using Simple JavaScript Web Based Games. *Journal of Engineering Education*, *90*(3), 347–355. doi:10.1002/j.2168-9830.2001.tb00613.x

Daunt, L. A., Umeonusulu, P. I., Gladman, J. R. F., Blundell, A. G., Conroy, S. P., & Gordon, A. L. (2013). Undergraduate teaching in geriatric medicine using computer-aided learning improves student performance in examinations. *Age and Ageing*, *42*(4), 541–544. doi:10.1093/ageing/aft061 PMID:23727555

Direito, I., Pereira, A., & Oliveira Duarte, A. M. (2012). Engineering Undergraduates' Perceptions of Soft Skills: Relations with Self-Efficacy and Learning Styles. *Social and Behavioral Science*, *55*, 843–851. doi:10.1016/j.sbspro.2012.09.571

Ditcher, A. K. (2001). Effective Teaching and Learning in Higher Education, with Particular Reference to the Undergraduate Education of Professional Engineers. *International Journal of Engineering Education*, *17*(1), 24–29.

Dym, C. L., Agogino, A. M., Eris, O., Frey, D. D., & Leifer, L. J. (2005). Engineering Design Thinking, Teaching and Learning. *Journal of Engineering Education*, *94*(1), 103–120. doi:10.1002/j.2168-9830.2005.tb00832.x

Eisner, E. W. (2000). *Benjamin Bloom 1913-99*. Retrieved from http://www.ibe.unesco.org/publications/ThinkersPdf/bloome.pdf

Ellis, G. W., Rudnitsky, A., & Silverstein, B. (2004). Using Concept Maps to Enhance Understanding in Engineering Education. *International Journal of Engineering Education*, *20*(6), 1012–1021.

Ellis, G. W., & Turner, W. A. (2003). Helping Students Organize and Retrieve Their Understanding of Dynamics. *Proceedings of the American Society for Engineering Education Annual Conference & Exposition*.

Errico, M., Maccioni, S., Tola, G., & Zuddas, P. (2007). A Deterministic Algorithm for the Synthesis of Maximum Energy Recovery Heat Exchange Network. *Computers & Chemical Engineering*, *31*(7), 773–781. doi:10.1016/j.compchemeng.2006.08.001

Evaluation Research Team. (2008). Using Graphs and Charts to Illustrate Quantitative Data. *Evaluation Briefs*, No. 12.

Felder, R. M., & Silverman, L. K. (1988). Learning and Teaching Styles in Engineering Education. *Engineering Education*, *78*(7), 674–681.

Freemantle, N., Wood, J., & Crawford, F. (1998). Evidence into practice, experimentation and quasi experimentation: Are the methods up to the task? *Journal of Epidemiology and Community Health*, *52*(2), 75–81. doi:10.1136/jech.52.2.75 PMID:9578853

Gologlu, C., & Zeyveli, M. (2009). A genetic approach to automate preliminary design of gear drives. *Computers & Industrial Engineering*, *57*(3), 1043–1051. doi:10.1016/j.cie.2009.04.006

Ho, C. F., Favero, J. P. D., & Tong, P. P. (2002). *Inexpensive Computer-Aided Learning Methods and Apparatus for Learners*. United States Patent.

Huang, Y. M., Liang, T. H., Su, Y. N., & Chen, N. S. (2012). Empowering personalized learning with an interactive e-book learning system for elementary school students. *Educational Technology Research and Development*, *60*(4), 703–722. doi:10.100711423-012-9237-6

Johnson, D. W., Johnson, R. T., & Smith, K. A. (1991). ASHE-ERIC Higher Education Report: Vol. 4. *Cooperative Learning Increasing College Faculty Instructional Productivity*. Washington, DC: The George Washington University, School of Education and Human Development.

Kampenes, V. B., Dyba, T., Hannay, J. E., & Sjoberg, D. I. K. (2009). A systematic review of quasi-experiments in software engineering. *Information and Software Technology*, *51*(1), 71–82. doi:10.1016/j.infsof.2008.04.006

Keyser, M. W. (2000). Active learning and cooperative learning: Understanding the difference and using both styles effectively. *Research Strategies*, *17*(1), 35–44. doi:10.1016/S0734-3310(00)00022-7

Linder, B. M. (1999). *Understanding Estimating and its Relation to Engineering Education* (Doctoral dissertation). Retrieved from DSpace @ MIT and Thesis Database.

Litzinger, T. A., Lattuca, L. R., Hadgraft, R. G., & Newstetter, W. C. (2011). Engineering Education and the Development of Expertise. *Journal of Engineering Education*, *100*(1), 123–150. doi:10.1002/j.2168-9830.2011.tb00006.x

Maisarah & Sidhu, M. S. (2013). An Algorithm to Simplify and Formulate Assumption in Gear Box Design Analysis. *Australian Journal of Basic and Applied Sciences*, *7*(10).

Maisarah. (2016). *An Approach to Enhance the Learning Curve of Malaysian Mechanical Engineering Undergraduates in Spur Gear Design* (Masters thesis). Universiti Tenaga Nasional (UNITEN).

Malekpour, A., Ling, T. C., & Lim, W. C. (2008). Location Determination Using Radio Frequency RSSI and Deterministic Algorithm. *IEEE Communication Network and Services Research Conference*, 488-495. 10.1109/CNSR.2008.32

Matousek, J., & Schwarzkopf, O. (1996). A Deterministic Algorithm for the Three-Dimensional Diameter Problem. *Computational Geometry*, *6*(4), 253–262.

Neumann, D. L., Neumann, M. M., & Hood, M. (2011). Evaluating Computer-Based Simulations, Multimedia and Animations that Help Integrate Blended Learning with Lectures in First Year Statistics. *Australasian Journal of Educational Technology*, *27*(2), 274–289. doi:10.14742/ajet.970

Padmanabhan, S., Ganesan, S., Chandrasekaran, M., & Raman, V. S. (2010). *Gear Pair Design Optimization by Genetic Algorithm and FEA. In Frontiers in Automobile and Mechanical Engineering* (pp. 301–307). FAME.

Perrotta, C., Featherstone, G., Aston, H., & Houghton, E. (2013). *Game-based Learning: Latest Evidence and Future Directions (NFER Research Programme: Innovation in Education)*. Slough: NFER.

Prince, M. (2004). Does Active Learning Work? A Review of the Research. *Journal of Engineering Education, 93*(3), 223–231. doi:10.1002/j.2168-9830.2004.tb00809.x

Raveendran, R., & Batmanabane, G. (2011). Alternatives to Animals in Teaching: Experience in an Indian Medical School. *Proceedings of the ALTEX 8th World Congress*, 355-360.

Reichardt, C. S. (2009). *The SAGE Handbook of Quantitative Methods in Psychology: Quasi-Experimental Design*. London, UK: SAGE Publications Ltd. doi:10.4135/9780857020994.n3

Sanchez, R. P., Bartel, C. M., Brown, E., & DeRosier, M. (2014). The acceptability and efficacy of an intelligent social tutoring system. *Computers & Education, 78*, 321–332. doi:10.1016/j.compedu.2014.06.013

Savsani, V., Rao, R. V., & Vakharia, D. P. (2010). Optimal weight design of a gear train using particle swarm optimization and simulated annealing algorithms. *Mechanism and Machine Theory, 45*(3), 531–541. doi:10.1016/j.mechmachtheory.2009.10.010

Schittek, M., Mattheos, N., Lyon, H. C., & Attstrom, R. (2001). Computer assisted learning. A Review. *European Journal of Dental Education, 5*(3), 93–100. doi:10.1034/j.1600-0579.2001.050301.x PMID:11520331

Serin, O. (2011). The Effects of the Computer-Based Instruction on the Achievement and Problem Solving Skills of the Science and Technology Students. *The Turkish Online Journal of Educational Technology, 10*(1), 183–201.

Shakerin, S. (2006). The Art of Estimation. *International Journal of Engineering Education, 22*(2), 273–278.

Sidhu, M. S. (2010). *Technology-Assisted Problem Solving for Engineering Education: Interactive Multimedia Applications*. Hershey, PA: Engineering Science Reference. doi:10.4018/978-1-60566-764-5.ch004

Sidhu, M. S., Selvanathan, N., Singh, D., & Ramesh, S. (2002). Current Challenges & Trends of Interactive Multimedia in Enhancing Engineering Education. *Proceedings of IEEE TENCON'02*, 1, 265-268. 10.1109/TENCON.2002.1181265

Sidhu. (2006). *Development of TAPS Packages for Engineering* (PhD Thesis). Universiti Malaya.

Smetana, L. K., & Bell, R. L. (2012). Computer Simulations to Support Science Instruction and Learning: A critical review of the literature. *International Journal of Science Education*, *34*(9), 1337–1370. doi:10.1080/09500693.2011.605182

Stach, W., Kurgan, L., & Pedrycz, W. (2010). A Divide And Conquer Method For Learning Large Fuzzy Cognitive Maps. *Fuzzy Sets and Systems*, *161*(19), 2515–2532. doi:10.1016/j.fss.2010.04.008

Stieger, J., Thelen, A. C., Bach, U., Richert, A., & Jeschke, S. (2014). Challenges of Large Group Study Courses in Engineering Education. *Automation. Communication and Cybernetics in Science and Engineering*, *2013/2014*, 503–511.

Vattam, S. S., Goel, A. K., Rugaber, S., Hmelo-Silver, C. E., Jordan, R., Gray, S., & Sinha, S. (2011). Understanding Complex Natural Systems by Articulating Structure-Behavior-Function Models. *Journal of Educational Technology & Society*, *14*(1), 66–81.

Yadav, A., & Barry, B. E. (2009). Using Case-based Instruction to Increase Ethical Understanding in Engineering: What Do We Know? What Do We Need? *International Journal of Engineering Education*, *25*(1), 138–143.

KEY TERMS AND DEFINITIONS

Deterministic Algorithm: In computer science, it is an algorithm which, given a particular input, will always produce the same output, with the underlying machine always passing through the same sequence of states.

Divide and Conquer Algorithm: Works by recursively breaking down a problem into two or more sub-problems of the same or related type, until these become simple enough to be solved directly. The solutions to the sub-problems are then combined to give a solution to the original problem.

Machinery Design: Serves innovators in mechanical design, design engineers and managers in OEM, processing, and R&D with technical content.

Quasi-Experiment: Is an empirical interventional study used to estimate the causal impact of an intervention on target population without random assignment. Quasi-experiments are subject to concerns regarding internal validity, because the treatment and control groups may not be comparable at baseline.

Spur Gear: Are the most common type of gears. They have straight teeth and are mounted on parallel shafts.

Chapter 6
Enhancing Multi-Body Mechanisms With Learning-Aided Cues in an Augmented Reality Environment

ABSTRACT

Augmented reality (AR) is a potential area of research for education, covering issues such as tracking and calibration, and realistic rendering of virtual objects. The ability to augment real world with virtual information has opened the possibility of using AR technology in areas such as education and training as well. In the domain of computer-aided learning (CAL), researchers have long been looking into enhancing the effectiveness of the teaching and learning process by providing cues that could assist learners to better comprehend the materials presented. Although a number of works were done looking into the effectiveness of learning-aided cues, none has really addressed this issue for AR-based learning solutions. This chapter discusses the design and model of an AR based software that uses visual cues to enhance the learning process and the outcome perception results of the cues.

INTRODUCTION

Computer simulations are providing unique insights into the way the world works today (Wolfram, 2002). Students can now experiment real problem

DOI: 10.4018/978-1-7998-0465-9.ch006

solving task in a virtual world of complex, dynamic systems in a way that was impossible before. Therefore numerous technologies have emerged and are being used in the educational sector to produce a better knowledge society. As such many technologies have emerged and are being used in the educational sector to enhance the learning process. Some of these technologies include the use of software, interactive white boards, multimedia and virtual reality. Most studies found in the literatures also reported the advantages and disadvantages of these technologies (Oleg, 2000); (John, et al. 2002); (Manjit Sidhu, 2009); (Manjit Sidhu, 2011). On the other hand the learning and understanding process of the information presented and learnt from the previously mentioned technologies has not improved much (Huber & Dietmajer 2010). A more recent technology i.e. augmented reality is becoming popular and being tested by many researchers to see if it could help learners visualize and understand the learning process better. Augmented Reality (AR) is commonly known as the ability to overlay computer graphics onto the real world. Unlike immersive Virtual Reality (VR), AR interfaces allow users to see the real world at the same time as virtual imagery attached to real locations and objects. AR concepts also use tangible user interface (TUI). Unlike physical interface whereby the user may interact with the software via a mouse or keyboard connection, TUI uses wireless interaction. Further details on how this technology works are given in (Liarokopis, 2002). The motivation of the research aims to look at alternative solutions to visualize engineering problems using AR and visual cues to enhance the learning process. The system developed in this research was used and tested by first year engineering students at UNITEN. User interviews and questionnaires were used to collect users understanding of the subject matter and perceptions of the learning aided cues employed. All of which showed positive response.

BACKGROUND

According to Bonnie Meyer in her book published in 1975, cueing means the addition of a non-content aspect of prose, which gives emphasis to certain aspects of the semantic content or points out aspects of the structure of the content (de Koning et al., 2010a). Hence, researchers have been identifying and looking at various important cues to aid learners in computer mediated learning environments. The use of multimedia elements in instructional design have also contributed to the importance of using suitable cues to aid in the delivery of multimedia enhanced contents (de Koning et al., 2010b); (Lijia et al., 2011). Appropriate cues in multimedia enhanced learning environment will reduce students cognitive load (Mayer & Moreno, 2003); Brianna et al., 2007). The advancements in the area of computer graphics and AR have resulted in the use of complex 2D and 3D synthetic computer generated objects in AR-based computer aided learning (CAL) applications. A number of researches have been done incorporating AR technology in various domains of educational and training scenarios (Zagoranski & Divjak, 2003); (Augmented Reality in Education, 2012). This project in particular, is looking into, not only using AR technology to assist in the delivery of mechanical engineering education, but also enhancing the learning process with suitable learning-aided cues.

CUES IN TEACHING AND LEARNING ENVIRONMENT

Researchers and developers have begun to realize the importance of cues in aiding learners to better comprehend contents presented in computer aided learning environments. Cues are meant to guide learners cognitive processing rather than providing new information (de Koning et al., 2010a). Visual cueing is the addition of non-content information (e.g., arrows, circles, and coloring) to visual representations. In multimedia based learning, learners are engaged in three important cognitive processes: Selecting, Organizing, and Integrating (Mayer & Moreno, 2003). The authors believe that the presence of cues is even more significant in learning environments where spatial intelligence and self-paced learning occurs, such as those in an AR-based learning environment. There are various types of cues which have been explored and researched in. Figure 1 provides (a rather crude and non-exhaustive) summary of these cues.

AUGMENTED REALITY IN ENGINEERING EDUCATION

Augmented Reality uses computer-driven technologies to deliver and augment sensory information to users, thus supplementing and/or replacing real-world sensations (Jerome, 2007). Milgram and Kisihino have described the relationship and taxonomy of AR technology along the Virtuality Continuum (VC) in a Mixed Reality (MR) environment (Lijia et al., 2011). Figure 2, shows the VC in which AR can be defined as a class of display where the

Figure 1. Types of learning aided cues

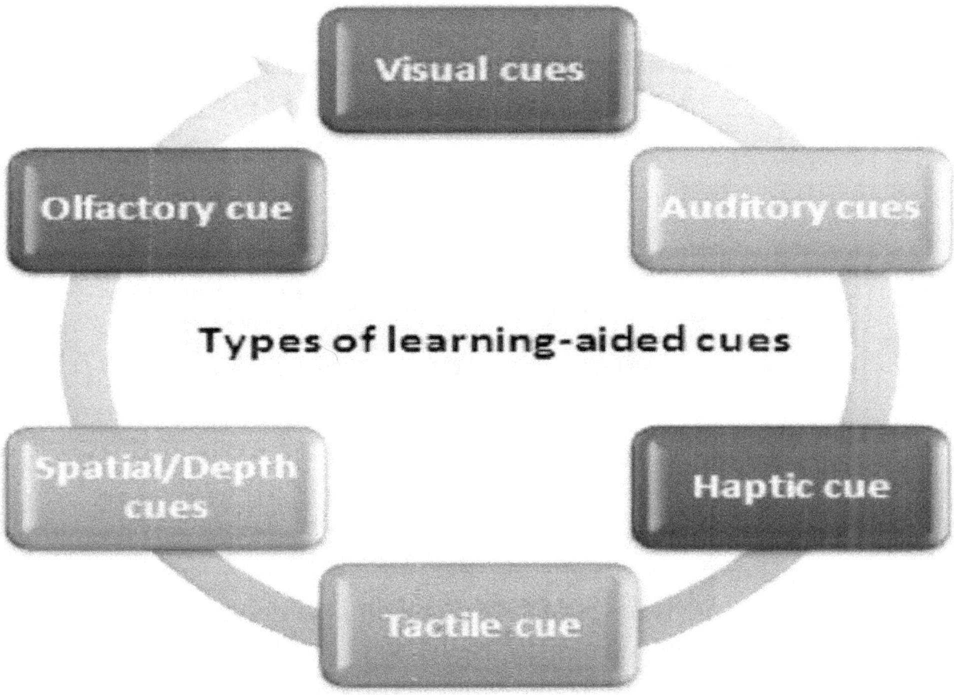

surrounding real environment is enhanced, overlaid or augmented with computer generated information. An Augmented Reality (AR) system can be defined as having these three properties (Azuma, et al., 2001).

- Combines real and virtual objects in a real environment,
- Runs interactively, and in real time; and
- Registers (aligns) real and virtual objects with each other.

According to Johnson, Levine, Smith and Stone, Augmented reality has strong potential to provide both powerful contextual, in situated learning experiences and serendipitous exploration and discovery of the connected nature of information in the real world (Johnson et al., (2010). As mentioned in the previous sections, AR has been used in delivering teaching and learning process. Some of the prominent and successful implementations of AR in the various teaching and learning domain, such as geometry, chemistry and biology are listed in (Zagoranski & Divjak, 2003); (Augmented Reality in Education, 2012). A major project where AR platform had been implemented

Figure 2. Reality-virtuality continuum (Milgram & Kishino, 1994).

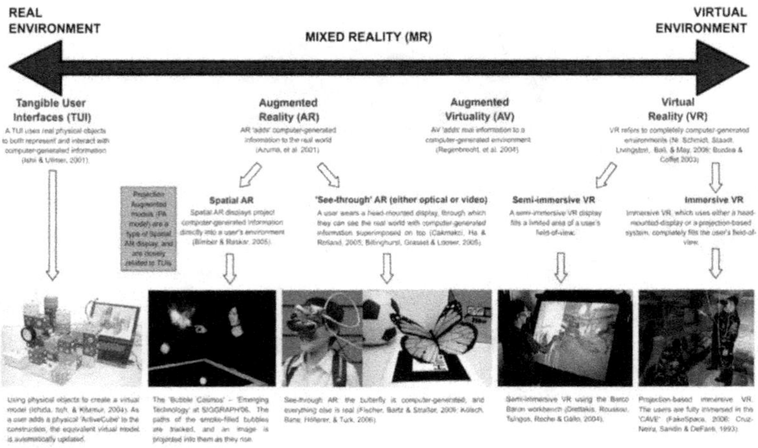

in school environment is the ARiSE project, where several European countries took part in this project (Balog et al., 2007).

One area where many have used AR as medium of delivering educational content is in the teaching of engineering syllabus. A learning support system based on Web3D and AR technologies have been developed to deliver contents related to engineering subjects such as platonic solids and machines (Liarokapis, & Mourkoussis, 2004).

An AR system for automotive engineering education has been developed with the objective of teaching disassemble/assemble procedures of an automatic transmission (Farkhatdinov, & Ryu, 2003). Another research looked into the possibility of using AR in the electrical and mechanical engineering education "*Technical Practice on Virtual reality and Augmented reality for Electrical and Mechanical engineering education*", (2010). In the research paper it was also argued that by combining AR technology with education software such as Matlab and Simulink will improve the pedagogy methods. A prototype Mixed Reality (MR) system where distributed CAVE-based workspaces have been developed to foster collaborative engineering between remote sites (Müller, 2007). AR-DEHAES, a learning toolkit for engineering graphics at tertiary level has been developed (Martin et al., 2010). This toolkit consists of a software application, an explanatory video, a notebook and an augmented book. Another recent project which is similar to the authors proposed project is an AR based simulation of multibody systems developed by (Valentini & Pezzuti, 2009). They use a fiducial/marker (Farkhatdinov, & Ryu, 2003), based

AR to augment real world with a virtual model of slider-crank mechanism using an AR video see-through system setup.

THEORETICAL BACKGROUND

The developed AR system simulates the workings of a multibody system for an engineering mechanics problem. This simulated multibody system is a dynamic behavior of interconnected rigid bodies, each of which may undergo translational and rotational displacements, see Figure 3. This type of a multibody system is known as the slider and crank mechanism. The slider-crank mechanism is an arrangement of mechanical parts designed to convert straight-line motion to rotary motion or vise-versa (Slider-crank, 2010). The AR system developed is based on the monitor based video displays or Window-on-the-world (WoW) display setup, see Figure 4. This set up uses merged video streams of virtual objects and real world s views, and displays these merged sequences of images on conventional computer monitors.

AUGMENTED REALITY IN AUGMENTED REALITY SYSTEM

In this study selected visual cues such as color, text or arrows, audio explaining concepts and animated mechanisms were implemented to see if this method could help students in understanding the mechanisms of an engineering problem presented in an AR environment. Since visual cues have not been used and tested in an augmented reality engineering application for learning, in this paper a 3D augmented reality (AR) application for teaching engineering was developed and tested. Seven patterns of visual/auditory sensory cues were embedded into the AR application to see if it could help students visualize an engineering problem i.e. slider crank mechanism. Sensory cues may include visual cues, auditory cues, tactile cues, haptic cues, olfactory cues, and others as mentioned earlier in Figure 1. Sensory cues play an important role in theories of perception, especially theories of appearance (how things look). The six visual cues (markers) (Figure 5) employed in this research for visualizing the slider crank problems are 1. Change color, 2. Display slider crank, 3. Display 2D diagram, 4. Display axes, 5. Display trail/path, and 6. Display all mechanisms. Only one auditory cue was used. The markers are used to allow the camera to calculate its respective orientation and distance, necessary for real time renderings. The system is composed of three components: tangible

Figure 3. The schematic of the multibody system, the slider and crank mechanism used in the proposed AR-based learning system

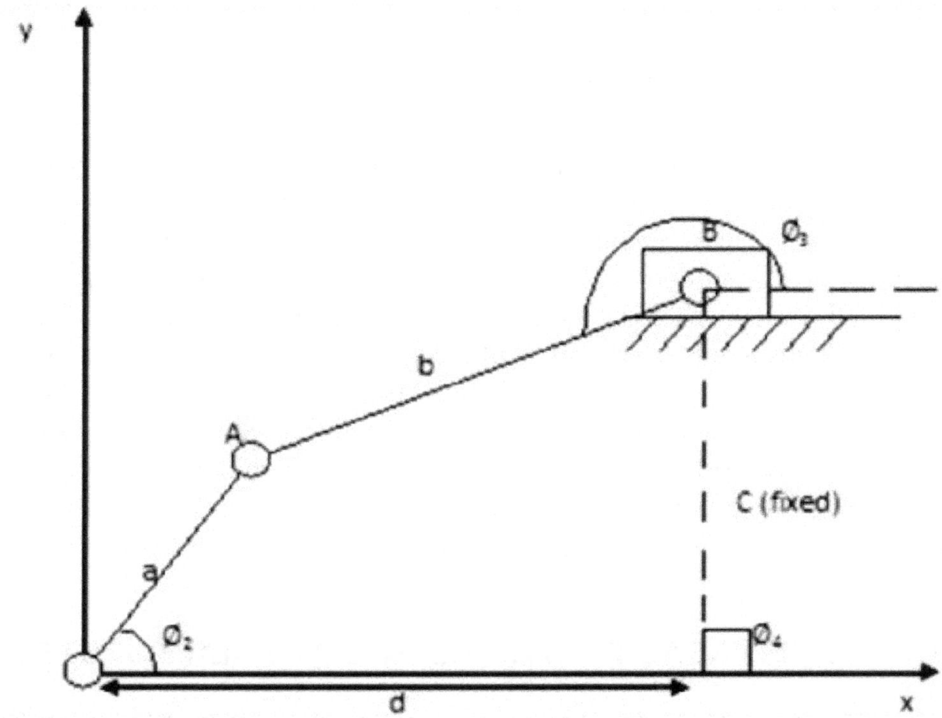

user interface, display component and rendering component. The tangible interface mimics a textbook problem based interface for user interaction. The

Figure 4. Monitor based AR system

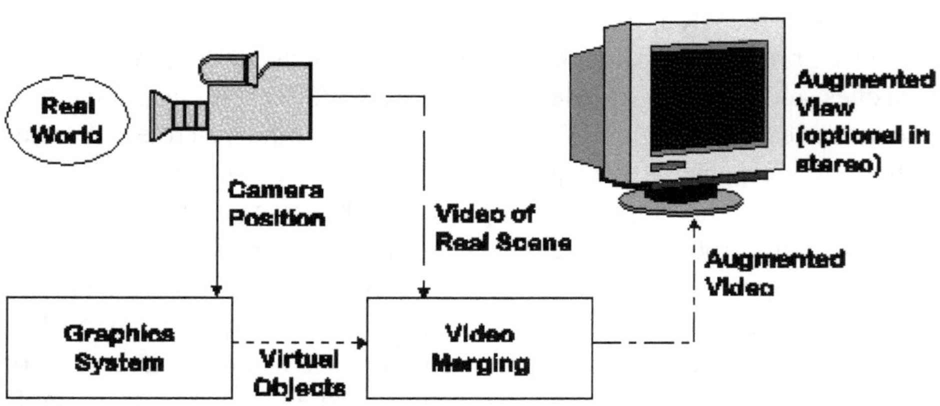

Figure 5. Markers to display the cues

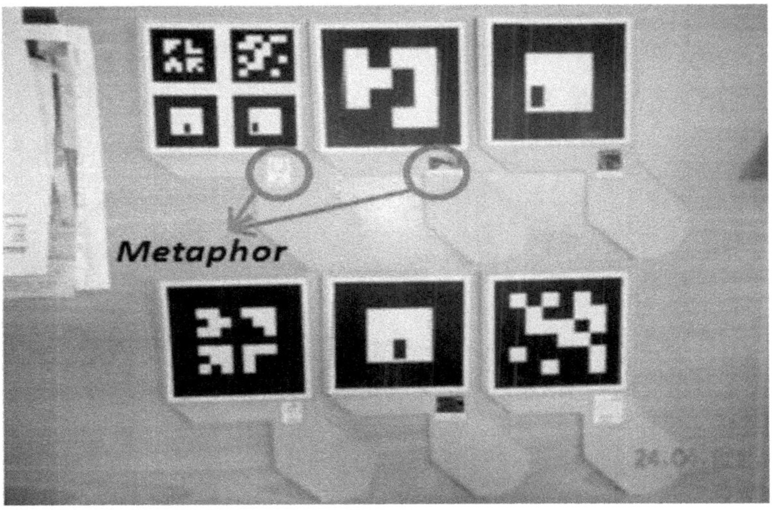

coordinate information of a particular cue is generated as soon as the camera recognizes the maker being shown and translated into the AR environment. The system setup is shown in Figure 6.

VISUALIZING THE ENGINEERING CONCEPTS USING VISUAL CUES

This section briefly explains the learning and visualization process of using the visual cues used in this study. As mentioned earlier in this chapter, six visual cues and one auditory cue were designed and used to see if students could recall and link the theory learnt earlier in the normal classroom. Each marker is used to perform a different task and display a cue. One of the cues used to display all the engineering mechanisms in the slider crank problem is shown in Figure 7. The system also narrates the user about the objectives and gives a brief explanation about the problem. Since not all users were familiar by interacting with the system using tangible user interface (TUI), we also designed the keyboard user interface that functions (alphabets used to perform a particular task) the same as the TUI. Examples of using visual cues are shown through Figures 8 and 9. Figure 8 depicts the cue to show the axes, Figure 9 depicts the cue to show/change the color of the slider crank. The student may also see multiple cues (two or more) by using two markers

Figure 6. The system setup

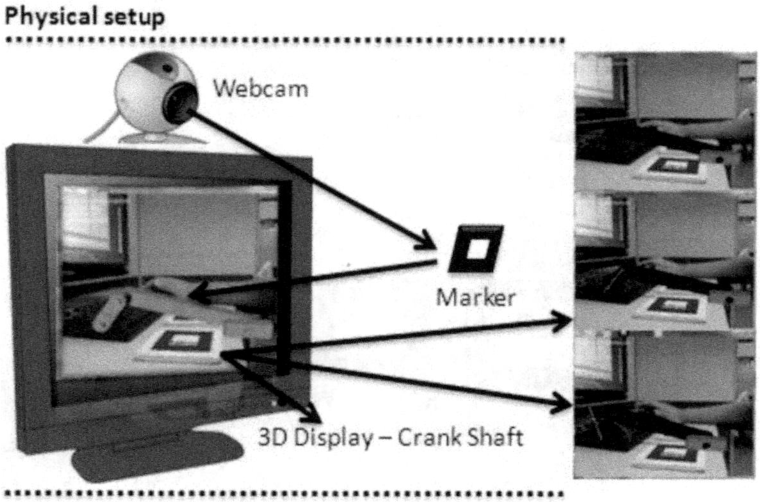

such as the one shown in Figure 10 (showing the motion trail and the 2D diagram of the slider crank mechanism).

RESULTS AND DISCUSSION

Education Environment and Evaluation

The AR system was tested with first year mechanical engineering students at UNITEN. A total of 37 students participated in the evaluation. Two systems were setup in the computer aided learning multimedia software (CALMUS) research laboratory. Each student spent approximately fifteen minutes in using the system. Interviews and questionnaires were used to measure the usefulness of the visual cues, how well does the system assist them in understanding the engineering concepts and effectiveness of the method in their learning.

Interviews and Questionnaires

To assess the system design, interviews and questionnaires, the most used evaluation techniques, were used to generate more in-depth data on users' visualization process and give designers insight into improving the system. An evaluation session was arranged after a lecture where each student was

Figure 7. All cues in the system

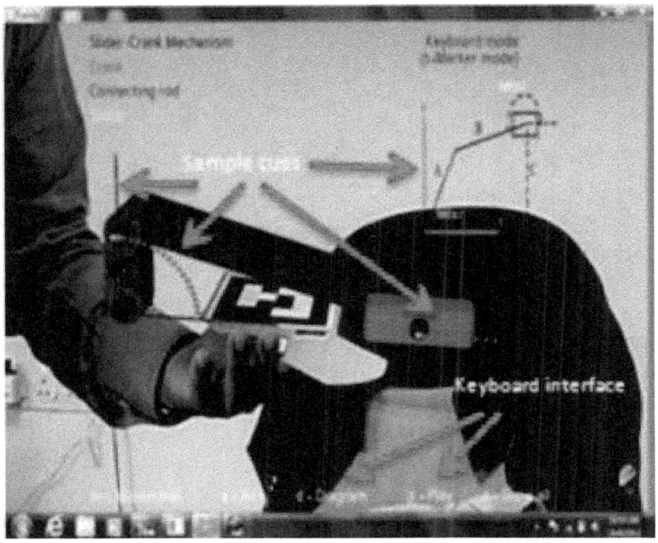

Figure 8. Single cue for showing the axes

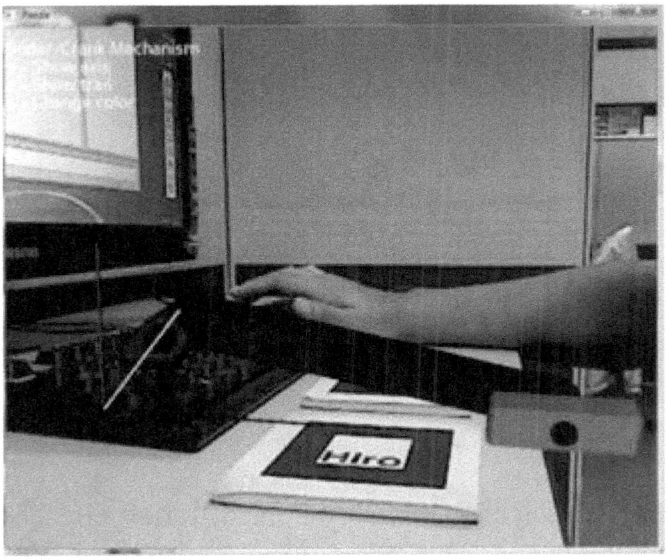

Figure 9. Cue showing the change of color

Figure 10. The motion trail and the 2D diagram

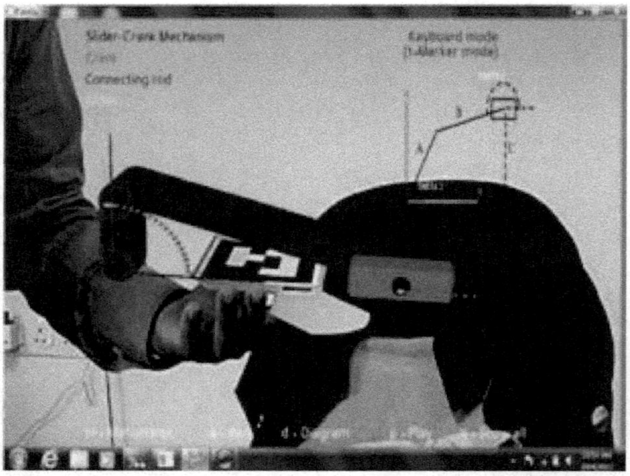

asked to elaborate their thoughts on the hands-on experience in order to identify usability concerns. They were also allowed to present their ideas and share any thoughts for improving the system. A survey was done to obtain user perceptions on the cues based on *Very useful*, *Useful* and *Not useful*.

Results of Interviews and Questionnaires

The results of the study showed that the weaker students (previously having problems in understanding the subject matter) who used the AR system in this study and had significantly better understanding on both comprehension and visualization of the problem. However some students had problems interacting with the tangible user interface as they were not familiar with the system. In general the students appeared to be satisfied by using the system and felt it helped them in their learning. However they also commented that the system was restricted by technology limitation when the AR system was difficult to be interacted with the markers as they had to hold them for some time. On the other hand they appreciated the use of 3D graphics and animation and the use of alphabets to interact with the system/display a particular visual cue. The students also revealed in the interviews that the AR system was more natural to interact with and looked very much like the real world object. The results in Figure 11, shows a positive impact on all the three aspects of using cues in this study. As a whole most students found all the cues to be Very useful and a small number of students found some of the cues were *not useful*. This could be a result of the students being unfamiliar of using tangible user interfaces and they had to spend more time adapting to this sort of user interface rather than concentrating on its contents using cueing method.

SUMMARY

Newer technologies are changing the way students learn today. In this study, a new system using augmented reality was developed and tested. Visual cues were employed to see if it could further enhance the learning of engineering problems. The learning aided cues helped students in reducing the cognitive workload when a single cue was used to display a particular engineering concept in the problem presented to them. The study found that the students benefited from using the system as they found it to be very useful and an interesting new method of learning. The technology limitations in this study are the future aims of the study.

Figure 11. Outcome perception of the cues

n=37

	Very useful	Useful	Not useful
All cues	16	14	7
Axes cue	25	9	3
Trail motion cue	34	2	1
Change color cue	12	16	9
Trail & 2D diagram cue	22	11	4
Auditory cue	17	14	6

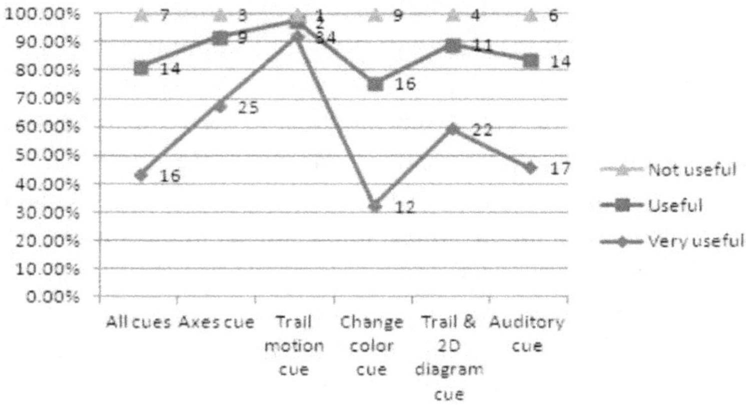

ACKNOWLEDGMENT

The author would like to acknowledge the significant contributions of Mohd Anwar Muslimen a GRA student who designed and developed the AR based Slider Crank ATAPS package.

REFERENCES

Augmented Reality in Education. (2012). Retrieved from http://wik.ed.uiuc.edu/index.php/Augmented_Reality_in_Education

Azuma, R., Baillot, Y., Behringer, R., Feiner, S., Julier, S., & MacIntyre, B. (2001). Recent Advances in Augmented Reality. *IEEE Computer Graphics and Applications*, *21*(6), 34–47. doi:10.1109/38.963459

Balog, A., Pribeanu, C., & Iordache, D. (2007). Augmented Reality in Schools: Preliminary Evaluation Results from a Summer School. *Proceedings of WASET International Conference on Technology and Education*, 114-117.

de Koning, B. B., Tabbers, H., Rikers, R. M. J. P., & Paas, F. (2010a). Attention Cueing as a Means to Enhance Learning from an Animation. *Applied Cognitive Psychology*, *21*(6), 731–746. doi:10.1002/acp.1346

de Koning, B. B., Tabbers, H. K., Rikers, R. M. J. P., & Paas, F. (2010b). Attention guidance in learning from a complex animation: Seeing is understanding? *Learning and Instruction*, *20*(2), 111–122. doi:10.1016/j.learninstruc.2009.02.010

Farkhatdinov, I., & Ryu, J. (2003). *Development of educational System for Automotive Engineering based on Augmented Reality*. Available from: http://robot.kut.ac.kr/papers/DeveEduVirtual.pdf

Huber, G., & Dietmajer, P. (2010). Four Bar: An educational synthesis and design tool for four bar mechanisms. *Proceedings of the 10th World Congress on the Theory Machines and Mechanisms*, 20-24.

Jerome, C. J. (2007). Effects of spatial and non-spatial multimodal cues on orienting of visual-spatial attention in an augmented environment. In *Technology-Based Training Research Unit*. Orlando, FL: U.S. Army Research Institute for the Behavioral and Social Sciences.

Johnson, L., Levine, A., Smith, R., & Stone, S. (2010). *The 2010 Horizon Report*. Austin, TX: The New Media Consortium.

Liarokapis, F., & Mourkoussis, N. (2004). Web3D and Augmented Reality to support Engineering Education. *World Transactions on Engineering and Technology Education*, *3*(1), 11–14.

Liarokopis, F. (2002). Multimedia Augmented Reality Interface for E-Learning (MARIE). *World Transactions on Engineering and Technology Education*, *1*(2).

Lin, L., & Robert, K. A. (2011). Using animations and visual cueing to support learning of scientific concepts and processes. *Computer Education*, *56*(3), 650–658. doi:10.1016/j.compedu.2010.10.007

Manjit Sidhu, S. (2009). Visualizing Curvilinear Cylindrical Engineering Members in 3-D Environment with Motion Path. In *NCM2009: 5th International Joint Conference on INC, IMS and IDC*. IEEE.

Manjit Sidhu, S. (2011). A 3-D Graphical Model for a Four Bar Linkage Mechanism Problem. In *First IRAST International Conference on Data Engineering and Internet Technology (DEIT)*. IEEE.

Mayer, R. E., & Moreno, R. (2003). Nine Ways to Reduce Cognitive Load in Multimedia Learning. *Educational Psychologist*, *38*(1), 43–52. doi:10.1207/S15326985EP3801_6

Müller, D. (2007). Mixed reality learning spaces for collaborative experimentation: A challenge for engineering education and training. *International Journal of Online Engineering*, *3*(4).

Oleg, V. (2000). *Fundamentals of Kinematics and Dynamics of Machines and Mechanisms*. CRC Press. DOI: doi:10.1201/9781420042337.ch1

Saorin, Morin, Contero, & Alcaniz. (2010). AR_Dehaes: An Educational Toolkit Based on Augmented Reality Technology for Learning Engineering Graphics. *Advanced Learning Technologies (ICALT), 2010 IEEE 10th International Conference on*, 133-137.

Scott, B. M., & Schwartz, N. H. (2007). Navigational spatial displays: The role of metacognition as cognitive load. *Learning and Instruction*, *17*(1), 89–105. doi:10.1016/j.learninstruc.2006.11.008

Slider-crank. (n.d.). Retrieved from http://www.britannica.com/EBchecked/topic/548729/slider-crank-mechanism

Judy, Vance, & Larochelle. (2002). Spatial Mechanism Design in Virtual Reality with Networking. *Journal of Mechanical Design*, *124*.

Technical Practice on Virtual reality and Augmented reality for Electrical and Mechanical engineering education. (2010). Retrieved from http://www.wseas.us/e-library/conferences/2008/venice/edu/edu04.pdf

Valentini, P., & Pezzuti, E. (2009). Interactive Multibody Simulation in Augmented Reality. *ECCOMAS Thematic Conference on Multibody Dynamics.*

Wolfram, S. (2002). *A New kind of science. Champaign, IL: Wolfram Media.*

Zagoranski, S., & Divjak, S. (2003). *Use of Augmented Reality in Education.* EUROCON Ljubljana. doi:10.1109/EURCON.2003.1248213

KEY TERMS AND DEFINITIONS

Markers: Are normal images or small objects which are trained beforehand so that they can be recognized later in the camera stream.

Multi-Body Mechanisms: A multibody dynamic (MBD) system is one that consists of solid bodies, or links, that are connected to each other by joints that restrict their relative motion. The study of MBD is the analysis of how mechanism systems move under the influence of forces, also known as forward dynamics.

Slider Crank: Linkage is a four-link mechanism with three revolute joints and one prismatic, or sliding, joint. The rotation of the crank drives the linear movement the slider, or the expansion of gases against a sliding piston in a cylinder can drive the rotation of the crank.

Visual Cues: Is a signal and reminder of something; aiming to be self–explanatory and pre-attentive, it brings to mind knowledge from previous experiences providing a framework for its own interpretation.

Chapter 7
Empirical Study Outcome of Augmented Reality Technology for Solving Engineering Problems in UNITEN

ABSTRACT

In this chapter, the preferences of UNITEN students in using augmented reality technology user interface applications to solve selected engineering problems are compared. Cross sectional study design and Wilcoxon-Signed Rank Test approach were adopted to analyze the difference in the rankings of the engineering applications. Within a controlled environment, this research further applies "trials within the same session but with breaks between tasks," an affirmed reliable method in measuring learnability that has been rarely explored by any related works locally. The results were captured through descriptive statistical analysis. The findings provided reliable evidences that multiple function user interface (MFIT) is more effective than the tangible user interface (TUI) for engineering students. Simultaneously, this research also presents evidences that MFIT is better than TUI in the engineering problem solving statistically.

DOI: 10.4018/978-1-7998-0465-9.ch007

INTRODUCTION

As defined in the literatures, "*AR is the digital overlay on top of the real world, consisting of computer graphics, text, video and audio, which is interactive in real time*" (Azuma, 1997; Billinghurst et al., 2014; Papagiannis, 2017). The technology can be interacted with and experienced through many electronic devices such as the computer, tablet, smartphone or AR eyewear equipped with the software and camera. In the educational and human computer interaction studies, many researchers have shown that AR can assist and improve learning as compared to traditional/conventional methods (Kerawalla et al., 2006; Chang et al., 2010; Lee, 2012). Although Lee (2012) stated that students may acquire knowledge and skills through different modes of instruction that include classroom lectures with textbooks and computers, AR is put to practice through different interfaces and the interaction between human and computers depends on computer interfaces. However, when using technology to address a particular educational material, user interaction is one of the highly debated areas. Many instructional designers in the past found that the user interface of a particular software hinders users from learning the contents effectively because the users are provided with too much of information and need to focus more on the interface rather than the contents (Jameson, 1995; Alpert et al., 2003; Ziefle & Bay, 2006; Georgiev & Georgieva, 2009; Wong, 2017). The same problem was also experienced by instructors and students in UNITEN to calculate and analyze the Four-Bar Linkage (4BL) mechanism problems because the problems are basically accomplished with diagrams, equations, charts, texts and formulas. The poor user interface makes it even more difficult for them to focus on the contents. Other researchers have also developed software tools to solve the 4BL mechanism problem but failed to engage the user in the learning and visualization process (Oleg, 2000; Kihonge et al., 2002; Manjit 2003; 2008; 2009; Huber and Dietmajer, 2010). In AR technology the two user interfaces debated in the literatures are the Tangible User Interface (TUI) interaction technique and the multiple function interaction technique (MFIT) as used in mobile applications, where the latter is frequently quoted to be an enhancement to TUIs disadvantages. In general both user interfaces have close relation to usability issues in the field of Human-Computer Interaction (HCI). Therefore, both are frequently compared and evaluated using usability measures. This research highlights the recent works done in the field of TUI vs. MFIT competitive study. Besides, this study was carried out to find research possibilities that could refine competitive usability evaluation by emphasizing on the enhancement of the new interaction technique, MFIT.

This research compares the preferences of UNITEN engineering students in using augmented reality user interfaces based four bar link (4BL) problem solving applications. The context of the study was a third-year mechanical engineering course in UNITEN, Malaysia.

PROBLEM STATEMENT

Reviews on several related works has raised some problem statements that have been determined in teaching and learning in engineering field especially mechanical engineering

In general, a few researches stated that the problems appeared among students in engineering education field (Ditcher, 2001; Linder & Flowers, 2001; Sidhu, et al., 2002; Ellis et al., 2004; Dym et al., 2005; Shakerin, 2006; Palmquist, 2007; Litzinger et al., 2011). From the student's view, it is difficult for students to interact and solve engineering problem with the present AR mechanism i.e. the users use markers (symbols that are recognized by the PC camera to overlay 3D objects on the display) to interact with the virtual model that is displayed on the screen. Thus, hinders users from visualizing the information. For example, in engineering problems many forms of information are presented to the students simultaneously such as text, charts, formulas/equations etc. which could cause difficulties for students to relate the information (Manjit, 2006; Yuen et al., 2011; Fang & Gou, 2016).

Besides this, there were problems that occurred among the lecturer (Ellis & Turner, 2003; Levy & Ben-Ari, 2007; Stieger et al., 2014) whereby that they need to repeat the problems to the students several times before the students can understand (Milgram et al., 1994; Maqableh & Manjit, 2013). For example, the lecturer may need to draw the diagrams and write the formula repeatedly until the students can understand.

Another problem is attributed to the existing software in UNITEN. This is due to poorly design software particularly the user interface and inability to access such software due to cost/licensing. In the present AR user interface, only one function can be executed at a time which limits the user to interact effectively and absorb the content (Manjit & Kather, 2013).

Learnability is a significant usability characteristics used to gauge how usable an interface is. However, very few TUI versus MFIT research works used this metric as major measurements. The few who used this metric, applied the simplest form of measurement rather than the most realistic form. None has applied the most realistic approach called “Trials within the Same Session but with Breaks between Tasks”, which is claimed to be one of the suitable method in gathering learnability data (Cheng, 2011). The aim of the research was to conduct an evaluation comparing TUI and MFIT based on users’ learning and visualization of the 4BL mechanism problem with the new interaction technique.

USER INTERFACES AND AUGMENTED REALITY

In general, user interfaces (UI) are used everywhere including software, UI is a mechanism to interact between human and machine. If the application does not have a good UI, it cannot reach the goal of the user. As such UI is very important in order to guide the users to reach the goals of the particular application. This work examines the UI and the primary focus is on UI particularly Augmented Reality (AR) UI. This section discusses the preliminary work and reviews on user interfaces of AR in different fields such as education and engineering with the AR interfaces and also performs a comparison between Tangible User Interface (TUI) and a new Multiple Function Interaction Technique (MFIT). The core areas of this research are User Experience (UX) and Human Computer Interaction (HCI). Augmented reality (AR) has become a useful technology and increased the popularity in various fields among the society. The term AR, is defined as the combination of real, physical environment with the additional of the virtual objects (2-D/3-D) (Singh & Singh, 2013) in order to provide or add more information from the object (Ariyana & Wuryandari, 2012). The enhancement of the user’s perception and the interaction with the real environment can be done by adding on the real environment with 3-D virtual objects that appears together in the same place as the real environment. This is the goal of developing an AR system (Azuma et al., 2001). AR has the potential to create high quality 3-D models that encourage users to deeply visualize into few areas especially in engineering, medical, education, science, entertainment and others. For example, in the field of engineering education, the modules are relied on the laboratory work during which the users have chances to apply the theory introduced in traditional material, e-book for visualization, diagnosis in medical field i.e. doctors can insert information on a patient’s medical record by using AR system (Ariyana & Wuryandari, 2012). Interactive AR tools are also useful

Figure 1. Reality virtuality (RV) continuum

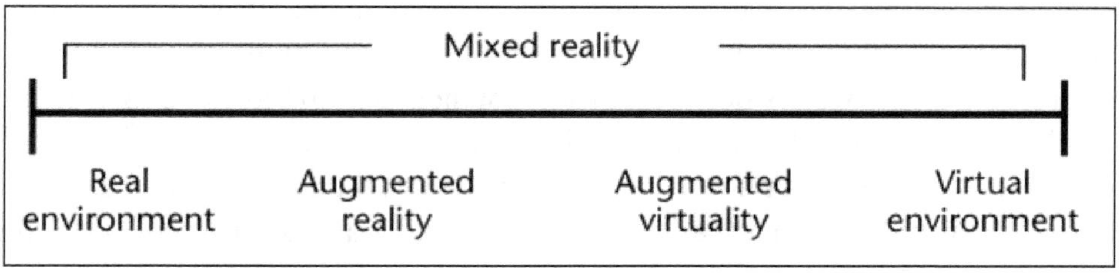

in visualizing the difficult problems by providing related advantages. One of the fundamental equipment associated with the AR technology is called the PC/Web camera/TUI/marker. Azuma's (2001) definition states that AR systems have the following three properties:

- The combination of virtual objects and real environment
- Real-time interaction, and
- Registered in three dimensions (3-D).

AR systems are defined as those which merge *"real and computer-generated information in a real environment, interactively and in real time, and virtual objects with physical ones"* (Yuen et al., 2011). Based on the Reality Virtuality Continuum, AR is closely connected to virtual reality (VR) since the concept of AR developed as an extension of VR (Ariyana & Wuryandari, 2012) as shown in Figure 1.

From the left of the continuum in Figure 1 defines as any environment which consists of real objects and includes whatever might be observed when viewing a real-world scene either through human or objects like window or displays (video). From right of the continuum defines as any environment which consists of virtual objects such as computer graphic simulations. Based on the framework, it defines that in which real world and virtual objects are put together within a single display, also means in anywhere between the extrema of the RV continuum is called a generic Mixed Reality (MR) (Milgram et al., 1994). There are two common Augmented Reality approaches namely (i) Marker-based and (ii) Marker-less AR that can show the information or data from databases or websites (Rose et al., 2010). The marked-based AR application is one of the AR approaches which will retrieve information or data from databases or direct users to, by using some particular logo or other graphic device. QR code is known as Quick Response code which is a 2-D bar code that can read by mobile phones equipped with suitable software. The marked-less AR is one of the AR approaches which involve the capability of Global Positioning System (GPS) of a smartphone. The GPS of smartphones has a high standard of positional accuracy (Rose et al., 2010). Users can use their smartphones' GPS to navigate to the directions. Since the existence of AR technology, many applications has been developed for general purpose by using the two interfaces (QR and GPS).

Although most of the interfaces designed are purposely at providing a better and natural control over AR applications, there are some problems when using the system (Mackey, 1998). In general, the dimension gap is one of the problems whereby AR is normally in three dimensional worlds. Sometimes six degrees of freedom (6DOF) is needed. For example, a touch-screen, camera or mouse only have X and Y coordinates while the interaction may need a Z coordinate which is the depth. In a sample situation, users are asked to grab a virtual object integrated to the scene captured by the camera or input numerical text for data analysis. In this situation, users normally hold markers so that the mobile device can track via the markers. But it seems difficult to grab the object using this method because the lack of depth with the camera. Another problem when using AR applications with a touch-screen, especially with mobile devices, the users normally has to put their hand or finger on the screen. This may cover the objects or other important elements. Once the objects or elements that the users want to manipulate are hidden, it is difficult for the users to move or adjust the position. On the other hand, it lacks haptic feedback (sense of touch) after using the particular AR application. For example, users try to grab a virtual object via camera view on the device either in 2-D or 3-D, in real world, the users' only grab in the air. This makes the users not having a feel to touch of grabbing an object they usually feel in real life.

AR is applied in the development for different types of applications across the industries which depend on the needs of the users. More users are exploring the AR field and they prefer convenient and easy applications which can ease or lighten their works (Mekni & Lemieux, 2014). However, user interfaces play important role to the user as human is the main factor whatever interaction occurs through the interface. The users do not look deeper into what interfaces are and they just want to know how good or how easy the interface works. There are a lot of user interaction techniques that have been used in augmented reality. Interaction technique such as motion to interact with digital media is prompt of the explosive adoption and growth of AR. There are many companies empowered to propose more natural and special virtual experiences for their audiences by launching their own versions of AR through motion capture games and interaction technologies. Tangible User Speech interfaces in general are used as a command-based interaction style where the users execute actions through verbal commands. Most of the researchers focus on how speech and gesture input can be used to manipulate virtual objects. In general, interaction techniques are an important issue in AR application. The interactions techniques can help in enhancing the current augmented reality applications to make them become better and natural. As mentioned earlier, this research aimed on developing and testing two AR based user interfaces namely TUI and MFIT. This interfaces are further explained next. Firstly the TUI which uses markers (symbols that are recognized by the PC camera to overlay 3D objects on the display) can be applied in AR interfaces with the enhanced display possibilities provided by virtual image overlays (Koelle et al., 2014). Secondly the multi-touch has been used in the devices with touch screens. Some of the devices still utilizes mouse and trackpads for navigation on a screen such as computer, mouse. Touch screens are implemented by letting the users to interact with the information on the display physically such as the PC or smart phones. Users can easily interact with the display by applying their actions and gestures of all kinds. Multi-touch user interface can also allow multiple users to interact in same system or applications at the same time. This action enables the collaboration among the interaction of multiple users and the system or applications. The flexibility of multi-touch can lead users to adapt the system or application easily. Therefore in the research we hypnotized that the multiple function interaction technique (MFIT) is better than single marker based interaction technique (TUI) for usability interaction.

FOUR-BAR LINKAGE (4BL) MECHANISMS

In this research, the 4BL mechanism problem was selected and designed using two AR user interface approaches. The aim was to compare the marker based (symbols that are recognized by the PC camera to overlay 3D objects on the display) and multiple function interaction technique (MFIT). 4BL mechanisms are also known as four-bar linkages which is the movable and simplest closed chain linkage. 4BL includes of four bodies, known as bars or links, which are connected in a loop by four joints (Beer & Johnston, 1997). The joints have various degrees of freedom (DOF) which allow motion between the links. More information on this is available on Wikipedia.

USABILITY TESTING

Usability is important in the development of a system or application. Firstly, users are one of the importance matters in usability because usability will be decided by the users. If the users do not prefer to use the usability, this will cause the work done to be useless. Good usability gives good impression to users and this makes users want to use back again but then, bad usability will only give problems for the users. Computers already become one of the technologies that provide users with more functions such as e-shopping, social network, latest news updating and it already become part of the daily lives. There are many different types of usability evaluation models such as Dix et. al. (1998), Nielsen (1993), ISO and other related models which are usually used for evaluating. According to Jakob Nielsen, he declares that usability consists of different components and he uses five aspects for usability model as follows (Wiberg, 2003):

- **Learnability***:* For users who use the system for the first time, the system should be easy to learn so that users can get some work done.
- *Efficiency:* Once the users get familiar with the system by learning it, a high level of productivity is possible.
- **Memorability***:* The system should be easy to remember. If the users want to return to the system after some period, they still able to use it without learning again.
- **Errors***:* The system should prevent errors and recover well from all possible errors made.
- **Satisfaction***:* The system should be pleasant to use. The users should have good impression and satisfied after experiencing the system.

Shneiderman, (2009) deduced seven usability testing measurements each with its own characteristics. In this research, Competitive Usability Testing (CUT) was used for the evaluation because this usability testing is mostly used for comparing a new interface design and the previous interface design. In a competitive usability testing, an expert runs a usability test on the design to gain insights on the usability work of other competitors. Therefore we found CUT to be a suitable testing approach which can be used to compare TUI and MFIT designs.

METHODS USED IN MEASURING LEARNABILITY AND EFFECTIVENESS

Tullis & Albert (2008) had defined that there are three ways that can be used to measure learnability (i) Trials within the same session, (ii) Trials within the same session but with breaks between tasks and (iii) Trials between sessions. Trials within the same session" is a method that needs participants to carry out the tasks, one right after the other. "Trials within the same session but with breaks between tasks" is another alternative where there will be a break between tasks but in the same day. Besides these, "Trials between sessions" is applied to participants who carry out the same tasks over multiple sessions and it may with at least one day in between (Tullis & Albert, 2008). In this research, "Trials within the same session but with breaks between tasks" was used to measure learnability and effectiveness. The short break can enable the students to fill in their questionnaires.

Study Duration, Design, Sampling

The duration of this study took about two years which started from May 2015 until May 2017. Since some data collection and monitoring the quality of the surveys may take a few years (Lehtonen et. al. 2004), we adopted the cross-sectional study design with samples from mechanical engineering (ME) course students from College of Engineering (COE), UNITEN.

Quantitative Data Collection

In this research, the software usability measurement Inventory (SUMI) by Kirakowski & Corbett, (1993) was used to measure the quality of the software depending on user's opinion. SUMI questionnaire is a method that can be used to measure the software quality from the end user's point of view. Sumi is also consistent to use for assessing the quality of the software or prototype. In SUMI questionnaires, it consists of scales of Agree, Undecided and Disagree. The original questionnaire has fifty questions with three components. However, in this research, only twenty questions related to the evaluation of TUI and MFIT were used. Students need to choose among three options which are (i) Agree, (ii) Average or (iii) Disagree accordingly for the 20 questions. In terms of observation, the covert observation method (involves the researcher joining the group he or she is studying) was used to observe the reaction of students

Figure 2. Comparison metaphors of TUI and MFIT design

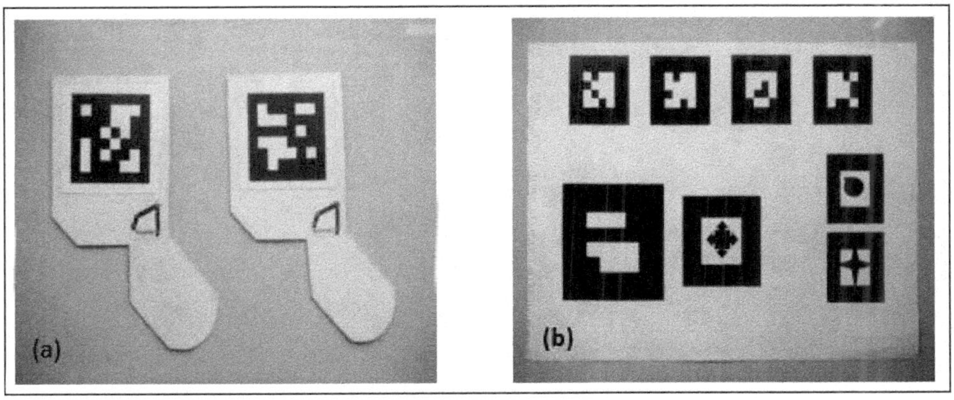

when they interacted with both the TUI and MFIT software interfaces. The observation took place when the students explored both the software and how they obtained the solutions for the 4BL problems by interacting with the software. The data was collected by writing down in notes according to the students' response to the software/interaction mechanisms.

Data Analysis

Data was analyzed using Statistical Package for the Social Sciences (SPSS) version 18.0. The descriptive analysis of categorical variables was done in Frequency and Percentage. Wilcoxon-Signed Rank Test was used to analyze the difference in the rankings between the TUI and MFIT.

Testing Process

The experiment was carried out in the computer aided learning, multimedia research (CALMUS) lab, College of Science Computer & Information Technology (CSIT), UNITEN. The duration of the experiment took half a day. The procedure was divided into two experiments testing stages with forty students. Figure 2(a) depicts the metaphor for the TUI user interface and Figure 2(b) depicts the metaphor for the MFIT user interface. Figure 3(a) shows the TUI marker-based 4BL application and Figure 3(a) & (b) shows MFIT based 4BL application being tested by the users.

Figure 3. (a) The TUI marker-based 4BL application; (b): MFIT based 4BL application

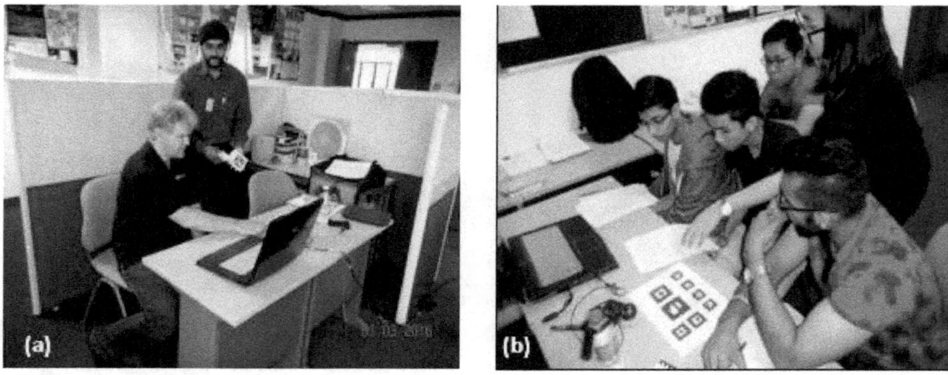

Figure 4. The TUI marker-based 4BL application with simulation and computations

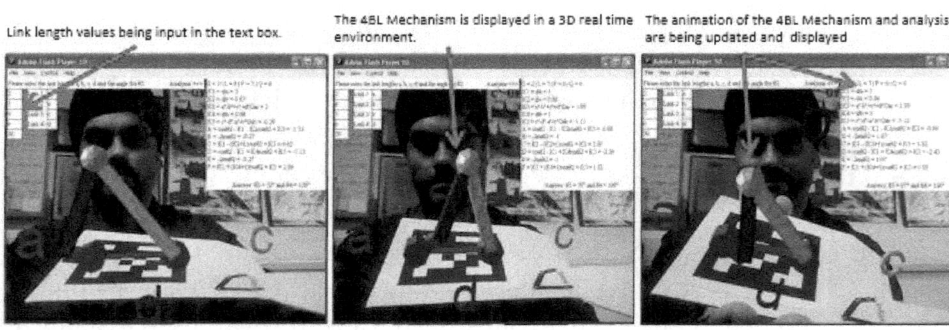

Figure 4 shows snapshots of the TUI marker-based 4BL application with simulation and computations of answers in real time environment and Figure 5 shows snapshots of the MFIT based 4BL application with simulation and computations.

RESULTS AND GENERAL DISCUSSION

The results from the feedback of forty students on the user interaction satisfaction are shown in Tables 1 – 3.

The results of SUMI questionnaires are combined using percent frequency values for clearer explanation and shown in Table 1. Based on Table 1, 21 students (52.5%) did not agree that the response of the MFIT software's inputs

Figure 5. The MFIT based 4BL application with simulations

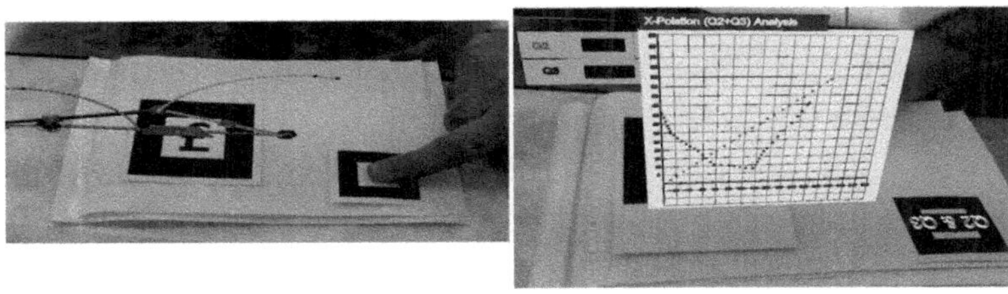

was too slow and 20 students (50%) agreed with the slow responds of the TUI software's inputs. There were 30 students (75.0%) of the respondents who would recommend this MFIT software to their friends and 19 students (47.5%) would not recommend the TUI software to their friends.

Besides these, the question on the helpfulness of the instructions and prompts were agreed by 32 students (80.0%) whereas 30 students (75.0%) did not agree that the instructions and prompts were helpful to them. There were 14 students (35.0%) who disagreed that first learning to operate the MFIT software was full of problems and 15 students (37.5%) agreed that learning to operate the TUI software initially was full of problems. Meanwhile, there were 17 students (42.5%) and 18 students (45%) who did not know what to do after when using the MFIT and TUI software respectively. The frequency of 35 students (87.5%) enjoyed the session when using the MFIT software and 31 students (77.5%) did not enjoy using the TUI software. Based on the results shown, 22 students (55.0%) did not agree that the help information provided by the MFIT software was not useful while for TUI software, 23 students (57.5%) still agreed that the information was not useful.

For both the MFIT and TUI software applications, most of the students could not decide whether the software was easy to restart once the problem was solved. A number of 22 students (55.0%) did not feel that it takes a long time to learn on how to use the MFIT software commands but 21 students (52.5%) agreed that the commands of the TUI software required longer time to learn. The satisfaction of 31 students (77.5%) was high when using the MFIT software and there were 28 students (70%) who were satisfied with the TUI software. Besides these, 32 students (80.0%) agreed that the MFIT software provided information with clear and understandable method but there were 27 students (67.5%) who disagreed with the TUI software. There were 22 students (55.0%) of the MFIT software and 20 students (50%) of the TUI software who could not decide whether the information was sufficient because they might not have enough time to explore all the level of the 4BL mechanisms. Among the respondents, there were 30 students (75.0%) who could understand and act on the information provided by the MFIT software but 25 students (62.5%) disagreed for TUI software. When using the MFIT software, 28 students (70.0%) could perform the tasks directly meanwhile when using the TUI software, 25 students (62.5%) disagreed with that.

On the other hand, The MFIT software helped 20 students (50.0%) to solve the problems they had by using the software and there were only 4 students (10%) who agreed that the TUI software was helpful. For learning new functions, 18 students (45.0%) could learn easily by using the MFIT software as compared to 19 students (47.5%) who felt that it was difficult to learn by using the TUI software. Meanwhile, 19 students (47.5%) felt that the MFIT software did not require too many steps to get the things to work but as for the TUI software, there were 22 students (55%) felt it requires many steps to work on. There were 22 students (55%) who could not decide whether the MFIT software did what they expected but there were 17 students (42.5%) agree that the TUI software did not do what they expected. For the MFIT software, 20 students (50%) disagreed that it was easy to forget on how to use the software meanwhile there were 20 students (50%) who could not decide when using the TUI software. For both software, there were 17 students (42.5%) and 20 students (50%) who agreed that they needed assistance most of the times when using the MFIT and TUI software respectively. The result for analysing the SUMI questionnaire using Wilcoxon Signed-Ranks Test is shown in Table 2.

a. M1 < T1 b. M1 > T1 c. M1 = T1 d. M2 < T2 e. M2 > T2 f. M2 = T2 g. M3 < T3 h. M3 > T3 i. M3 = T3 j. M4 < T4 k. M4 > T4 l. M4 = T4 m. M5 < T5 n. M5 > T5 o. M5 = T5 p. M6 < T6 q. M6 > T6 r. M6 = T6 s. M7 < T7 t. M7 > T7 u. M7 = T7 v. M8 < T8 w. M8 > T8 x. M8 = T8 y. M9 < T9 z. M9 > T9 aa. M9 = T9 ab. M10 < T10 ac. M10 > T10 ad. M10 = T10 ae. M11 < T11 af. M11 > T11 ag. M11 = T11 ah. M12 < T12 ai. M12 > T12 aj. M12 = T12 ak. M13 < T13 al. M13 > T13 am. M13 = T13 an. M14 < T14 ao. M14 > T14 ap.M14 = T14 aq. M15 < T15 ar. M15 > T15 as. M15 = T15 at. M16 < T16 au. M16 > T16 av. M16 = T16 aw. M17 < T17 ax. M17 > T17 ay. M17 = T17 az. M18 < T18 ba. M18 > T18 bb.M18 = T18 bc. M19 < T19 bd. M19 > T19 be. M19 = T19 bf. M20 < T20 bg. M20 > T20 bh. M20 = T20

According to the result shown in the Test Statistics (Table 3), there are three questions which are having p-value less than 0.05. The results for

Table 1. SUMI questionnaires (TUI and MFIT) (% Frequency)

	Statements	TUI			MFIT		
		Agree	Average	Disagree	Agree	Average	Disagree
1.	This software responds too slowly to inputs.	20 (50.0)	9 (22.5)	11 (27.5)	6 (15.0)	13 (32.5)	21 (52.5)
2.	I would recommend this software to my friends.	9 (22.5)	12 (30.0)	19 (47.5)	30 (75.0)	10 (25.0)	0
3.	The instructions and prompts are helpful.	3 (7.5)	7 (17.5)	30 (75.0)	32 (80.0)	8 (20.0)	0
4.	Learning to operate this software initially is full of problems.	15 (37.5)	13 (32.5)	12 (30.0)	13 (32.5)	13 (32.5)	14 (35.0)
5.	I sometimes don't know what to do next with this software.	18 (45.0)	12 (30.0)	10 (25.0)	17 (42.5)	15 (37.5)	8 (20.0)
6.	I enjoy my sessions with this software.	0	9 (22.5)	31 (77.5)	35 (87.5)	5 (12.5)	0
7.	I find that the help information given by this software is not very useful.	23 (57.5)	10 (25.0)	7 (17.5)	6 (15.0)	12 (30.0)	22 (55.0)
8.	If this software stops, it is not easy to restart it.	10 (25.0)	30 (75.0)	0	0	31 (77.5)	9 (22.5)
9.	It takes too long to learn the software commands.	21 (52.5)	11 (27.5)	8 (2.5)	8 (20.0)	10 (25.0)	22 (55.0)
10.	Working with this software is satisfying.	2 (5.0)	10 (25.0)	28 (70.0)	31 (77.5)	8 (2.5)	1 (2.5)
11.	The way that system information is presented is clear and understandable.	0	13 (32.5)	28 (67.5)	32 (80.0)	8 (20.0)	0
12.	There is never enough information on the screen when it's needed.	15 (37.5)	20 (50.0)	5 (12.5)	4 (10.0)	22 (55.0)	14 (35.0)
13.	I can understand and act on the information provided by this software.	5 (12.5)	10 (25.0)	25 (62.5)	30 (75.0)	8 (20.0)	2 (5.0)
14.	Tasks can be performed in a straightforward manner using this software.	6 (15.0)	9 (22.5)	25 (62.5)	28 (70.0)	10 (25.0)	2 (5.0)
15.	The software has helped me overcome any problems I have had in using it.	4 (10.0)	13 (32.5)	23 (57.5)	20 (50.0)	16 (40.0)	4 (10.0)
16.	Learning how to use new functions is difficult.	19 (47.5)	13 (32.5)	8 (20.0)	10 (25.0)	13 (32.5)	19 (47.5)
17.	There are too many steps required to get something to work.	22 (55.0)	10 (25.0)	8 (20.0)	8 (20.0)	13 (32.5)	19 (47.5)
18.	The software hasn't always done what I was expecting.	17 (42.5)	15 (37.5)	8 (20.0)	4 (10.0)	22 (55.0)	14 (35.0)
19.	It is easy to forget how to do things with this software.	12 (30.0)	20 (50.0)	8 (20.0)	4 (10.0)	16 (40.0)	20 (50.0)
20.	I have to look for assistance most times when I use this software.	20 (50.0)	13 (32.5)	7 (17.5)	17 (42.5)	12 (30.0)	11 (27.5)

Table 2. Wilcoxon signed rank test (SUMI questionnaire)

Ranks				
		N	**Mean Rank**	**Sum of Ranks**
M1. This software responds too slowly to inputs. - T1. This software responds too slowly to inputs.	Negative Ranks	4[a]	7.50	30.00
	Positive Ranks	19[b]	12.95	246.00
	Ties	17[c]		
	Total	40		
M2. I would recommend this software to my friends. - T2. I would recommend this software to my friends.	Negative Ranks	28[d]	15.23	426.50
	Positive Ranks	1[e]	8.50	8.50
	Ties	11[f]		
	Total	40		
M3. The instructions and prompts are helpful. - T3. The instructions and prompts are helpful.	Negative Ranks	35[g]	18.00	630.00
	Positive Ranks	0[h]	.00	.00
	Ties	5[i]		
	Total	40		
M4. Learning to operate this software initially is full of problems. - T4. Learning to operate this software initially is full of problems.	Negative Ranks	13[j]	15.08	196.00
	Positive Ranks	16[k]	14.94	239.00
	Ties	11[l]		
	Total	40		
M5. I sometimes don't know what to do next with this software. - T5. I sometimes don't know what to do next with this software.	Negative Ranks	13[m]	14.00	182.00
	Positive Ranks	13[n]	13.00	169.00
	Ties	14[o]		
	Total	40		
M6. I enjoy my sessions with this software. - T6. I enjoy my sessions with this software.	Negative Ranks	40[p]	20.50	820.00
	Positive Ranks	0[q]	.00	.00
	Ties	0[r]		
	Total	40		
M7. I find that the help information given by this software is not very useful. - T7. I find that the help information given by this software is not very useful.	Negative Ranks	3[s]	8.50	25.50
	Positive Ranks	24[t]	14.69	352.50
	Ties	13[u]		
	Total	40		
M8. If this software stops, it is not easy to restart it. - T8. If this software stops, it is not easy to restart it.	Negative Ranks	0[v]	.00	.00
	Positive Ranks	17[w]	9.00	153.00
	Ties	23[x]		
	Total	40		

continued on following page

Table 2. Continued

Ranks				
		N	Mean Rank	Sum of Ranks
M9. It takes too long to learn the software commands. - T9. It takes too long to learn the software commands.	Negative Ranks	7[y]	13.43	94.00
	Positive Ranks	24[z]	16.75	402.00
	Ties	9[aa]		
	Total	40		
M10. Working with this software is satisfying. - T10. Working with this software is satisfying.	Negative Ranks	34[ab]	17.50	595.00
	Positive Ranks	0[ac]	.00	.00
	Ties	6[ad]		
	Total	40		
M11. The way that system information is presented is clear and understandable. - T11. The way that system information is presented is clear and understandable.	Negative Ranks	37[ae]	19.00	703.00
	Positive Ranks	0[af]	.00	.00
	Ties	3[ag]		
	Total	40		
M12. There is never enough information on the screen when it's needed. - T12. There is never enough information on the screen when it's needed.	Negative Ranks	4[ah]	14.75	59.00
	Positive Ranks	22[ai]	13.27	292.00
	Ties	14[aj]		
	Total	40		
M13. I can understand and act on the information provided by this software. - T13. I can understand and act on the information provided by this software.	Negative Ranks	33[ak]	18.52	611.00
	Positive Ranks	2[al]	9.50	19.00
	Ties	5[am]		
	Total	40		
M14. Tasks can be performed in a straightforward manner using this software. - T14. Tasks can be performed in a straightforward manner using this software.	Negative Ranks	30[an]	16.60	498.00
	Positive Ranks	2[ao]	15.00	30.00
	Ties	8[ap]		
	Total	40		
M15. The software has helped me overcome any problems I have had in using it. - T15. The software has helped me overcome any problems I have had in using it.	Negative Ranks	27[aq]	16.89	456.00
	Positive Ranks	4[ar]	10.00	40.00
	Ties	9[as]		
	Total	40		
M16. Learning how to use new functions is difficult. - T16. Learning how to use new functions is difficult.	Negative Ranks	6[at]	12.00	72.00
	Positive Ranks	18[au]	12.67	228.00
	Ties	16[av]		
	Total	40		

continued on following page

Table 2. Continued

Ranks				
		N	Mean Rank	Sum of Ranks
M17. There are too many steps required to get something to work. - T17. There are too many steps required to get something to work.	Negative Ranks	7[aw]	13.29	93.00
	Positive Ranks	23[ax]	16.17	372.00
	Ties	10[ay]		
	Total	40		
M18. The software hasn't always done what I was expecting. - T18. The software hasn't always done what I was expecting.	Negative Ranks	7[az]	13.00	91.00
	Positive Ranks	21[ba]	15.00	315.00
	Ties	12[bb]		
	Total	40		
M19. It is easy to forget how to do things with this software. - T19. It is easy to forget how to do things with this software.	Negative Ranks	6[bc]	10.50	63.00
	Positive Ranks	20[bd]	14.40	288.00
	Ties	14[be]		
	Total	40		
M20. I have to look for assistance most times when I use this software. - T20. I have to look for assistance most times when I use this software.	Negative Ranks	11[bf]	9.09	100.00
	Positive Ranks	12[bg]	14.67	176.00
	Ties	17[bh]		
	Total	40		

the three questions are Question 4 ($Z = -0.482$, $p = 0.630$), Question 5 ($Z = -0.172$, $p = 0.864$) and Question 20 ($Z = -1.193$, $p = 0.233$). Although these three questions have p-value which is less than 0.05, they still showed the result that MFIT was better than TUI. As for Question 4, there are 16 positive ranks that showed that learning to operate the MFIT initially has fewer problems if compared to TUI. For Question 20, there are 12 positive ranks that showed that the student does not need to look for assistance when using the MFIT interface.

As a result, the Wilcoxon Signed-Rank Test shows that the observed difference of 17 questions out of 20 questions between both the scores is significant ($p < 0.05$). This means that the null hypothesis is rejected and the alternative hypothesis is accepted. As such there is a difference between MFIT and TUI user interfaces. The overall ranks for MFIT are higher than those for TUI as shown in Table 2. We can therefore conclude that MFIT user interface is better than TUI user interface.

Table 3. Test Statistics table for SUMI questionnaire

Software	Z	Asymp. Sig. (2-tailed)
M1 – T1	-3.400[a]	.001
M2 – T2	-4.664[b]	.000
M3 – T3	-5.392[b]	.000
M4 – T4	-0.482[a]	.630
M5 – T5	-0.172[b]	.864
M6 – T6	-5.734[b]	.000
M7 – T7	-4.062[a]	.000
M8 – T8	-3.945[a]	.000
M9 – T9	-3.114[a]	.002
M10 – T10	-5.289[b]	.000
M11 – T11	-5.488[b]	.000
M12 – T12	-3.197[a]	.001
M13 – T13	-5.000[b]	.000
M14 – T14	-4.527[b]	.000
M15 – T15	-4.223[b]	.000
M16 – T16	-2.298[a]	.022
M17 – T17	-2.963[a]	.003
M18 – T18	-2.694[a]	.007
M19 – T19	-3.029[a]	.002
M20 – T20	-1.193[a]	.233

Note: T is question for TUI & M* is question for MFIT.* **Test Statistics**[c]
a. Based on negative ranks
b. Based on positive ranks
c. Wilcoxon Signed Ranks Test

CONCLUSION

By conducting an analysis with a sample of forty students among the mechanical engineering students from COE, UNITEN, this study presents a comparison research between the single marker based or (TUI) and multiple function interaction technique (MFIT) for evaluating the user interface and usability.

Firstly, this research looked into the problems that are faced by the students and lecturers in understanding the four-bar linkage (4BL) mechanism problem and also the existing 4BL applications. Although there were some existing 4BL applications, those applications were found to be misleading or unsuitable to be used by students in their learning due to some erroneous results.

In this research, the single marker based or (TUI) was noted as one of the 4BL AR application that is used to visualize the 4BL problems in UNITEN because there was a problem of using the application as it could only allow one function to be executed at one time. Due to this, students had difficulties in concentrating on the contents and visualizing the problem. Therefore, the Multiple Function Interaction Technique (MFIT) which was one of the recent applications developed in UNITEN, was tested and compared with the single marker based or (TUI) to determine if the MFIT is better than the TUI. The study revealed that the MFIT design approach was the preferred user interface as the students could concentrate more on the contents while interacting with the 4BL problem. The MFIT interface did not deviate their attention as they felt it was similar to using a natural keyboard. The Wilcoxon-Signed Rank Test further provided evidence that the overall reaction to the MFIT interface was positive in terms of terminology and system interaction, screen, learning and system capabilities. The study suggest that when interacting with problem solving software, the user should be free from cognitive workload whereby the user should fully focus and concentrate on the problems shown on the screen. In such problem solving software, the user may need to apply the right equations, values, view schematic diagrams, 2D/3D simulations and read useful information.

ACKNOWLEDGMENT

The author would like to acknowledge the significant contributions of Jee Geak Ying a GRA Masters in ICT student who did the comparative study of the AR applications, and statistically analyzed the data for the research.

REFERENCES

Alpert, S. R., Karat, J., Karat, C.-M., Brodie, C., & Vergo, J. G. (2003). User Attitudes Regarding a User-Adaptive eCommerce Web Site. *User Modeling and User-Adapted Interaction*, *13*(4), 373–396. doi:10.1023/A:1026201108015

Ariyana, Y., & Wuryandari, A. I. (2012). Virtual Interaction at Virtual Environment applied for Augmented Reality. *2012 International Conference on Cloud Computing and Social Networking (ICCCSN)*. 10.1109/ICCCSN.2012.6215734

Azuma, R., Baillot, Y., Behringer, R., Feiner, S., Julier, S., & MacIntyre, B. (2001). Recent Advances in Augmented Reality. *IEEE Computer Graphics and Applications*, *21*(6), 34–47. doi:10.1109/38.963459

Azuma, R. T. (1997). A Survey of Augmented Reality. *Journal Presence: Teleoperators and Virtual Environments*, *6*(4), 355–385. doi:10.1162/pres.1997.6.4.355

Beer, P., & Johnson, R. (1997). *Vector Mechanics for Engineers (Dynamics)*. McGraw-Hill Book.

Billinghurst, M., Clark, A., & Lee, G. (2015). A Survey of Augmented Reality. *Journal Foundations and Trends in Human-Computer Interaction*, *8*(2-3), 73–272. doi:10.1561/1100000049

Chang, G., Morreale, P., & Medicherla, P. (2010). Applications of augmented reality systems in education. In D.Gibson & B. Dodge (Eds.), *Proceedings of Society for Information Technology & Teacher Education International Conference 2010*, (pp. 1380-1385). Academic Press.

Cheng. (2011). *Advances in Multimedia Modeling: 17th International Multimedia Conference*. Springer.

Chin, J. P., Diehl, V. A., & Norman, K. L. (1988). Development of an Instrument Measuring User Satisfaction of the Human-Computer Interface. ACM CHI' 88 Proceedings, 213-218. doi:10.1145/57167.57203

Ditcher, A. K. (2001). Effective Teaching and Learning in Higher Education, with Particular Reference to the Undergraduate Education of Professional Engineers. *International Journal of Engineering Education*, *17*(1), 24–29.

Dix, A., Finlay, J., Abowd, G. D., & Beale, R. (1998). *Human-Computer Interaction* (2nd ed.). Prentice-Hall, Inc.

Dym, C. L., Agogino, A. M., Eris, O., Frey, D. D., & Leifer, L. J. (2005). Engineering Design Thinking, Teaching and Learning. *Journal of Engineering Education*, *94*(1), 103–120. doi:10.1002/j.2168-9830.2005.tb00832.x

Ellis, G. W., Rudnitsky, A., & Silverstein, B. (2004). Using Concept Maps to Enhance Understanding in Engineering Education. *International Journal of Engineering Education*, *20*(6), 1012–1021.

Ellis, G. W., & Turner, W. A. (2003). Helping Students Organize and Retrieve Their Understanding of Dynamics. *Proceedings of the 2003 American Society for Engineering Education Annual Conference & Exposition.*

Fang, N., & Guo, Y. (2016). Interactive computer simulation and animation for improving student learning of particle kinetics. *Journal of Computer Assisted Learning, 32*(5), 443–455. doi:10.1111/jcal.12145

Georgiev, T. S., & Georgieva, E. (2009). User Interface Design for Mobile Learning Applications. *International Conference on e-Learning and the Knowledge Society – e-Learning'09.*

Huber, G., & Dietmajer, P. (2010). Four Bar: An educational synthesis and design tool for four bar mechanisms. *Proceedings of the 10th World Congress on the Theory Machines and Mechanisms*, 20-24.

Jameson, A. (1995). Numerical uncertainty management in user and student modeling: An overview of systems and issues. *User Modeling and User-Adapted Interaction, 5*(3-4), 193–251. doi:10.1007/BF01126111

Kerawalla, L., Luckin, R., Seljeflot, S., & Woolard, A. (2006). Making it real: Exploring the potential of augmented reality for teaching primary school science. *Virtual Reality (Waltham Cross), 10*(3-4), 163–174. doi:10.100710055-006-0036-4

Kihonge, N., Pierre, L., & Judy, M. (2002). Spatial Mechanism Design in Virtual Reality with Networking. *Journal of Mechanical Design, 124*(3), 435–440. doi:10.1115/1.1481363

Kirakowski, J., & Corbett, M. (1993). SUMI: The Software Usability Measurement Inventory. *British Journal of Educational Technology, 24*(3), 210–212. doi:10.1111/j.1467-8535.1993.tb00076.x

Koelle, M., Lindemann, P., Stockinger, T., & Kranz, M. (2014). *Human-Computer Interaction with Augmented Reality* (Report no. 2198-9494). Embedded Interactive Systems Laboratory, University of Passau.

Lee, K. (2012). The Future of Learning and Training in Augmented Reality. *InSight: A Journal of Scholarly Teaching, 7*, 31-42.

Lehtonen, R., & Pahkinen, E. (2004). *Practical methods for Design and Analysis of Complex Surveys*. John Wiley and Sons.

Levy, R., & Ben-Ari, M. (2007). We Work So Hard and They Don't Use It: Acceptance of Software Tools by Teachers. *SIGCSE Bulletin*, *39*(3), 246–250. doi:10.1145/1269900.1268856

Linder, B., & Flowers, W. C. (2001). Integrating Engineering Science and Design: A Definition and Discussion. *International Journal of Engineering Education*, *17*(4), 436–439.

Litzinger, T. A., Lattuca, L. R., Hadgraft, R. G., & Newstetter, W. C. (2011). Engineering Education and the Development of Expertise. *Journal of Engineering Education*, *100*(1), 123–150. doi:10.1002/j.2168-9830.2011.tb00006.x

Mackey, W. E. (1998). Augmented Reality: Linking real and virtual worlds a new paradigm for interacting with computers. *Proceedings of the working conference on advanced visual interfaces*, 13-21. 10.1145/948496.948498

Manjit, S. S. (2008). Comparison of TAPS Packages for Engineering. *International Campus-Wide Information Systems*, *25*(3), 161–175. doi:10.1108/10650740810886330

Manjit, S. S. (2009). *Visualizing Curvilinear Cylindrical Engineering Members in 3-D Environment with Motion Path.* Fifth International Joint Conference on INC, IMS and IDC (NCM 2009). Seoul, Korea.

Manjit, S. S., & Kather, A. (2013). Computer Aided Simulation of a 4BL Engineering Problem Using Matlab. *Australian Journal of Basic and Applied Sciences*, *7*(10), 381–387.

Manjit, S. S., Ramesh, S., Selvananthan, N., & Diljit, S. (2003). Multimedia dilemma: choosing the right tool for authoring and teaching engineering. *World Conference on Science and Technology Education (ICASE 2003)*, Penang, Malaysia.

Maqableh, W., & Sidhu, M. (2013). *Interactive Four-Bar Linkage 3D-Simulation Using Augmented Reality.* The 6th International Conference on Information Technology.

Mekni, M., & Lemieux, A. (2014). Augmented Reality: Applications, Challenges and Future Trends. *Proceedings of the 13th International Conference on Applied Computer and Applied Computational Science (ACACOS '14)*, 205-214.

Milgram, P., Takemura, H., Utsumi, A., & Kishino, F. (1994). Augmented Reality: A class of displays on the reality-virtuality continuum. *Proceedings of the Telemanipulator and Telepresence Technologies, 2351*, 282–292. doi:10.1117/12.197321

Nielsen, J. (1993). *Usability Engineering*. San Francisco, CA: Morgan Kaufmann Publishers Inc. doi:10.1016/B978-0-08-052029-2.50007-3

Oleg, V. (2000). *Fundamentals of Kinematics and Dynamics of Machines and Mechanisms*. CRC Press.

Palmquist, S. (2007). *Active Project-Based Learning in Structural Analysis: Field Inspection of a Steel Truss Bridge*. Paper presented at the American Society for Engineering Education, Honolulu, HI.

Papagiannis, H. (2017). *Augmented Human: How Technology is shaping The New Reality*. O'Reilly Media.

Rose, S., Potter, D., & Newcombe, M. (2010). *Augmented Reality: A Review of available Augmented Reality packages and evaluation of their potential use in an educational context. In Unlocking the Hidden Curriculum*. University of Exeter.

Shakerin, S. (2006). The Art of Estimation. *International Journal of Engineering Education, 22*(2), 273–278.

Shneiderman, B. (2009). *Designing the User Interface: Strategies for Effective Human-Computer Interaction*. Addison-Wesley Publishing Book.

Sidhu, M. S., Selvanathan, N., Singh, D., & Ramesh, S. (2002). Current Challenges & Trends of Interactive Multimedia in Enhancing Engineering Education. *Proceedings of IEEE TENCON'02*, 1, 265-268. 10.1109/TENCON.2002.1181265

Singh, M., & Singh, M. P. (2013). *Augmented Reality Interfaces*. North Caroline State University. IEEE Computer Society. doi:10.1109/MIC.2013.107

Stieger, J., Thelen, A. C., Bach, U., Richert, A., & Jeschke, S. (2014). *Challenges of Large Group Study Courses in Engineering Education*. Automation. *Communication and Cybernetics in Science and Engineering, 2013/2014*, 503–511.

Tullis, T., & Albert, B. (2008). *Measuring the User Experience: Collecting, Analyzing, and Presenting Usability Metrics*. San Francisco, CA: Morgan Kaufmann Publishers Inc.

Wiberg, C. (2003). *A Measure of Fun Extending the scope of web usability* (Report no. 1401-4572). Umeå: Umeåuniversitet.

Wong, E. (2017). *How to Prevent Emotions in the User Experience of Your Product.* Retrieved from https://www.interaction-design.org/literature/article/how-to-prevent-negative-emotions-in-the-user-experience-of-your-product

Yuen, S., Yaoyuneyong, G., & Johnson, E. (2011). Augmented Reality: An Overview and Five Directions for AR in Education. *Journal of Educational Technology Development and Exchange*, *4*(1), 11–140. doi:10.18785/jetde.0401.10

Ziefle, M., & Bay, S. (2006). How to overcome disorientation in mobile phone menus a comparison of two different types of navigation aids. *International Journal of Human-Computer Interaction*, *21*(4), 393–433. doi:10.120715327051hci2104_2

KEY TERMS AND DEFINITIONS

Competitive Usability Testing: A kind of usability test on the design and competitors.

Global Positioning System: It is a global navigation satellite system that provides geolocation and time information to a GPS receiver anywhere on or near the Earth.

Metaphor: Is a figure of speech that describes an object or action in a way that isn't literally true, but helps explain an idea or make a comparison.

MFIT: Multiple function interaction technique that allows multiple task to be performed simultaneously (i.e., using two or more fingers to select options on the screen, selecting, dragging, resizing image on the screen, etc.).

Mixed Reality (MR): Sometimes referred to as hybrid reality, is the merging of real and virtual worlds to produce new environments and visualizations where physical and digital objects co-exist and interact in real time.

TUI: Is a user interface in which a person interacts with digital information through the physical environment.

Wilcoxon-Signed Rank Test: Is a non-parametric statistical hypothesis test used to compare two related samples, matched samples, or repeated measurements on a single sample to assess whether their population mean ranks differ (i.e., it is a paired difference test).

Chapter 8
Advanced Augmented Reality TAPS Software for Visualizing 4BL Mechanisms with Touch to Print Technique

ABSTRACT

Many learning methods have changed the way students learn. One method that is achieving much attention is augmented reality (AR). AR is a technology that blends simulated and real environment during the learning, interaction, and visualization process. As such, an AR ATAPS with a new interaction technique (touch-to-print) was designed, tested, and evaluated. The aim was to provide an improved user interface i.e. without having to use markers so as the learner could focus more on the visualization process. The AR ATAPS is capable of recognizing the 4BL mechanisms (based on Grashof's law and user input data) (i.e., drag-link, crank-rocker, double-rocker, and parallelogram linkage) and have been used in this study as an adjunct to traditional problem-solving method. The touch to print interaction technique, which is the main contribution of this research, has been useful to engage the user in the problem with a new interactive and learning experience as compared to the previous method (i.e., the use of markers to interact with a virtual object). The interaction technique method uses seven functions that are recognizable by the ATAPS (rotational, link colour change A, B, C, D, pause and voice command) for the user to touch the symbols on the paper and the system to model and analyze accordingly in real-time 3D environment. This study explores how far AR technology has come to support students in their

DOI: 10.4018/978-1-7998-0465-9.ch008

learning and interest in using this technology. The objective of this chapter was to determine the usefulness of touch to print interaction user interface for an AR application. A hands-on practical lab was conducted with first year engineering students at UNITEN. The evaluation and effectiveness of the AR ATAPS as an alternative to textbooks and current software learning packages was examined by means of a single-institutional evaluation study using mainly statistical quantitative techniques and ANOVA analysis. The prospective study (total sample size = 30) at University Tenaga Nasional (UNITEN) validated aspects of AR ATAPS interaction technique and provided feedback on the interface design and its problem-solving method. The results of the study showed that most of the participants never been experienced with AR applications before, but the ideas of implementing AR as a simulation tool for learning the kinesthetic and dynamic subjects is well accepted with a very beneficial feedback. Based on the findings, it was found that there is a positive changing in terms of the visualizing and imagining of the four-bar linkage mechanisms (4BL) which led to a good understanding of this subject. Further development of AR applications in the learning environment is being discussed.

INTRODUCTION

In general AR is a computer based system which has the ability to combine the real world and computer generated data. In this system, virtual objects are blended into real scene in the real time. Due to this capability, instructors can imagine the high potential that this technology might have if employed in the field of education. Ludwig et. al., (2005) define AR as Human-computer-interaction, which adds virtual objects to real senses that are provided by a video camera in real time. On the other hand Zhou et. al., (2008), simply defines AR as a technology "*which allows computer generated virtual imagery to exactly overlay physical objects in real time*". Others define AR as a system that combines the real world with the computer-generated information in a real environment, interactively and in real time, and which align virtual objects with physical ones, (Höllerer et. al., 2004).

Generally, AR combines three dimensional (3D) computer-generated objects and text superimposed onto real images and video all in real time. An interesting definition of AR has been described by (Azuma, 1997), as a variation of virtual reality (VR). VR technology completely immerses a user inside a synthetic environment. While immersed, the user may not see the surrounding real world. It has been argued in an essay that: *"Simulation of reality is not enforced by augmented reality. Instead, it takes a real object or space as the foundation and incorporates technologies that add contextual data to deepen a person's understanding of the subject*" (EduCause, 2005). This means the human perception is augmented or supplied with information not ordinarily detectable by human senses. An AR system can be identified by three properties which are the real and virtual objects are combined in a real environment, secondly, it runs interactively and in real time and thirdly, AR also aligns real and virtual objects with each other. In the learning environment, AR allows learners to interact with the real and virtual environments and manipulate objects that are not real. In addition to learning tasks and skills that have the ability to engage a reader in ways that never been possible.

Since the early 1990's, many researchers have been working on how to implement AR for educational use (Fitzmaurice, 1993). "Within the field of Education, AR applications have to be grounded in sound pedagogy to justify their cost and not be viewed as a passing gimmick or latest shiny object that is intriguing" these were the comments made by Sterling (2010). The educational AR is more of the theoretical research than the practical application till now. It has also been stated that the previous hindrances of cost and software are now diminishing because of the availability of Smartphone and Apps (Sterling, 2010).

The use of AR applications in learning can improve the spatial abilities of students when learning science subjects such as math and geometric objects (Durlach et. al., 2000). In learning chemical subject AR allows the students to inspect molecules from multiple viewpoints, controlling the interactions of molecules and it can increase the understanding of the concepts of chemical molecules (Maier, 2009).

According to Manfred (2006), AR can be integrated as a teaching platform into the teachers and students environment for displaying audio-visual and multimedia content in line with the needs in the relevant areas. In addition AR can be used in training to provide rich contextual learning for individuals learning skill; Sarnoff Corporation demonstrated the first augmented reality training system for US War fighters (Henderson, 2007).

It has been pointed out that the potential power of AR as a learning tool is that it can improve the students' ability to see the world around them in new ways and engage with realistic issues in a context with which the students are already connected (Erkan et. al., 2010).

The prime features of AR are visualization, realistic and interactivity. As such features like viewing 3D objects in real time and improving visualization, could play a significant role in the learning process and move the learning methods to the next level.

This Chapter presents an AR ATAPS with a new interaction technique (touch-to-print) was designed, tested and evaluated. The aim was to provide an improved user interface i.e. without having to use markers so as the learner could focus more on the visualization process. The AR ATAPS is capable of recognizing the 4BL mechanisms (based on Grashof's law and user input data) i.e. drag-link, crank-rocker, double-rocker and parallelogram linkage and have been used in this study as an adjunct to traditional problem solving method. The touch to print interaction technique which is the main contribution of this research, has been, useful to engage the user in the problem with a new interactive and learning experience as compared to the previous method i.e. the use of markers to interact with a virtual object.

PROBLEM STATEMENT

The vision of the college of engineering (COE) UNITEN is to emerge as 'A leading engineering college in energy and sustainability' with a mission to 'Advancing engineering knowledge through education and research innovation with industrial linkages' (COE, 2013).

Electronics and Communication Engineering, Electrical Power Engineering, Mechanical Engineering, Civil Engineering and an Engineering Mathematics & Management Unit are the four main departments of COE (UNITEN, 2013). The college offers programs at all Bachelor, Masters and PhD levels.

The main focus in this research will be on problems encountered by students of Mechanical Engineering during the course of Engineering Mechanics Dynamics. The instructors of the subject from their personal evaluation have noted that students found it difficult to understand the visualizations of dynamic models and their complex equations. To judge the problem accurately and to solve the problem to some extent is the main reason for choosing the subject for this research.

Manjit (2007) has stated that lack of availability of modern teaching equipment and methods in an institution may result in many problems. For example, what is taught during the lecture is not completely absorbed and understood by the students.

At the undergraduate learning one expects a variety of new teaching methods to be used in the class rooms, especially for Mechanical Engineering courses. In general, at the undergraduate level teaching Mechanics Dynamics course is often limited to the use of algebraic and trigonometry or to the use of some CAL packages which are mentioned earlier. They do not serve the nature of these courses and it needs special thinking and vocalization skills. As a result some students remain unable to understand the subject matter and require extra effort both from teachers and themselves to evolve from the crisis (Manjit, 2007).

To address the difficulties posed when teaching Mechanics Dynamics course, the proposed methods should be implemented to facilitate solving the engineering problems. This study is proposed to employ and use multimedia and AR technologies to help students visualize and understand the selected engineering problem(s).

However, the aim is that the proposed technique outlined in this study will assist instructors in teaching Mechanical Engineering courses, as well as the students in visualizing and learning.

MOTIVATION OF STUDY

In the perspective of Mechanical Engineering the following issues have motivated this study:

1. It was observed from engineering students who had studied Mechanical Engineering course, difficulties in understanding the concepts of the Linkages mechanisms unit, which affect grades in that course. This was a prime motivation for this research to evaluate and see if computer based learning aids such as ATAPS software could help engineering students to analyze, visualize and understand how to solve such engineering problems more efficiently.
2. Development of new learning/teaching software tools based on the new technologies such as Augmented Reality (AR) for engineering education

Figure 1. Video see-through (Marchese, 2008)

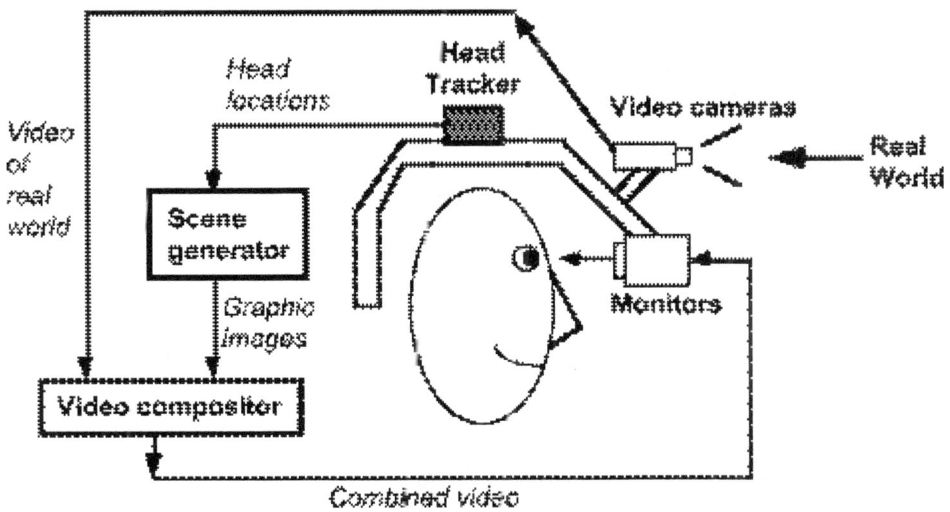

Figure 2. Optical see-through (Marchese, 2008)

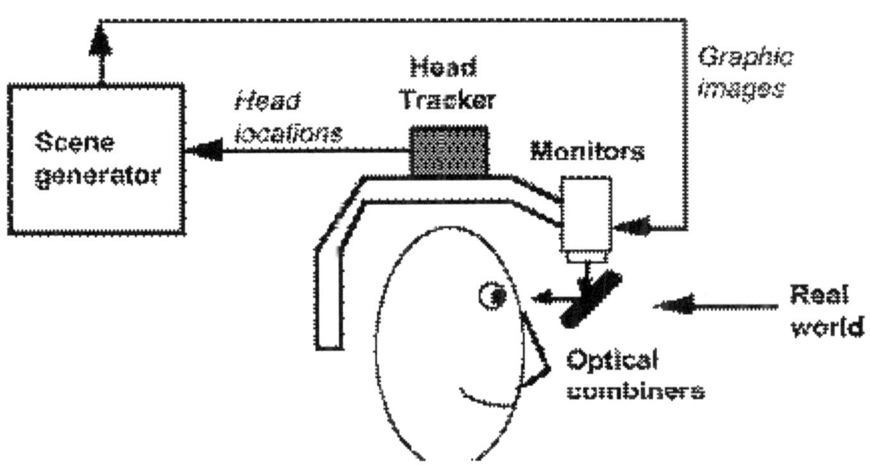

Figure 3. Testing of the ATAPS Software with HMD

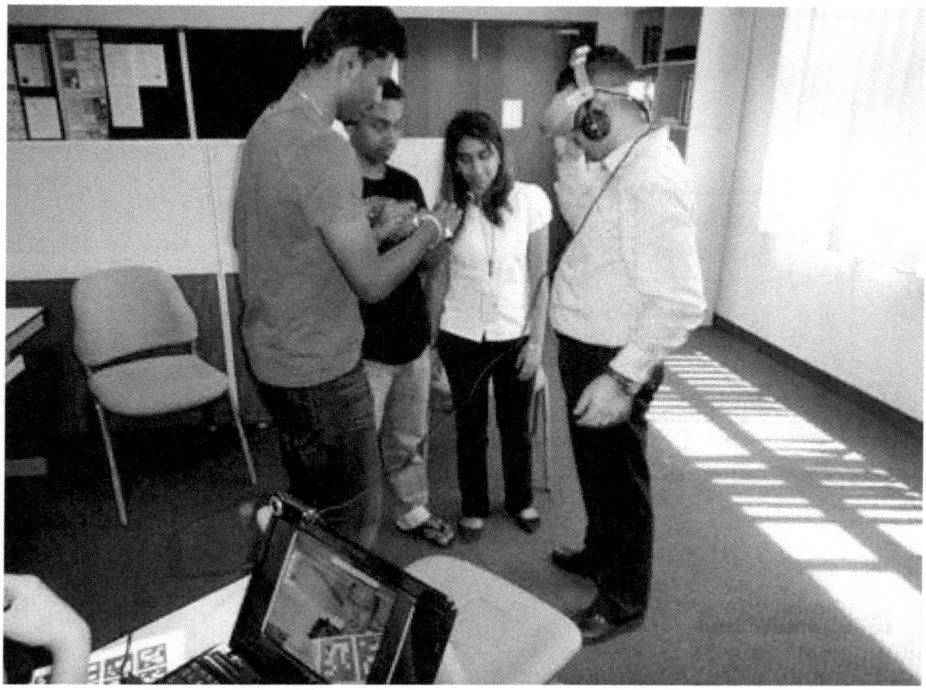

is appreciated in general but there is a need to take/perform practical initiatives.

Manjit (2007) stressed that:

"Education is a lifelong pursuit and involves observable, scientific, and empirical evidence to improve its' educational offerings".

By referring to the statement above, it has been assumed in this study that in order to understand engineering principles in a better way, ATAPS software should be frequently used. It is hoped that usage of ATAPS software would

Figure 4. Displays projector and & handheld projector

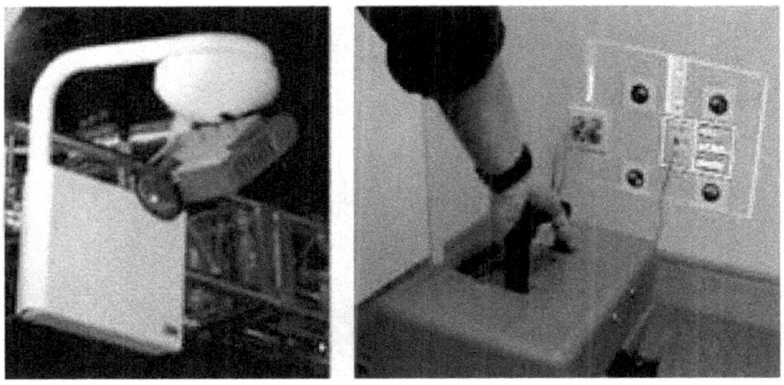

facilitate students in practical application of the principles taught during the lecture.

CHARACTERISTICS OF AUGMENTED REALITY

The use of AR is not limited on adding objects to the real environment. It has the ability and the potential to remove them by hiding these parts from the graphic overlay. When the individual must complete a particular task they eradicate all the other elements from the real world and focus all their attention on this mission. AR is not just limited to visual information, but it also has a connection with sound and haptic that can be used to change a perspective of the real world. Most of these elements were considered to be significant to be used in the ATAPS software.

Display Systems

To accomplish the augmentation task, there are two approaches that can combine the real and virtual world (Optical and Video technologies). Considering a see-through head mounted display (HMD) it entails that the user has the ability to view the real world that is present in front of them (Figure 1). It is the optical blending method that utilizes the open view HMD (Figure 2). With this method the particular user will be able to view the real world through HMD that is combined with the head mounted video cameras.

When the HMD does not show an image it will indicate that the user have a blind vision but still there is also a possibility that the user will be able to see the real world, but may have a feeling that is similar to wearing heavy glasses. Both of these methods can be accomplished by looking through the

Figure 5. Spatially aware handheld displays

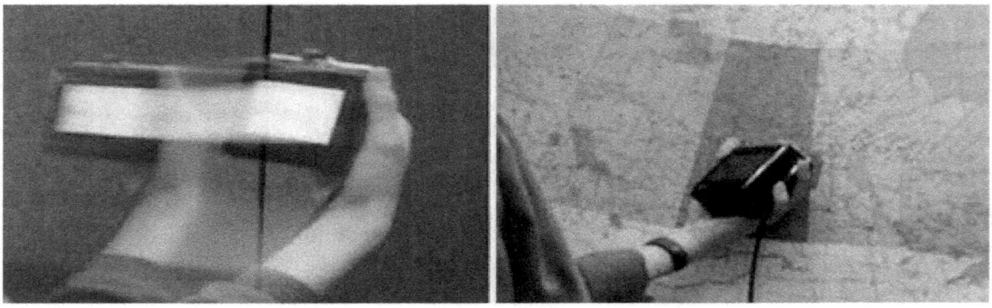

HMD. The ATAPS software was also tested with potential users by using the HMD see Figure 3.

However, according to Rolland (1994), it should be kept in mind that the optical case has a close connection with the intuitive definition of the whole see through concept.

Both of the optical and video technologies have advantages. As an example, the simplicity in the optical approach does not require blending the video streams and there is no need to worry about any delay that may occur between the two streams. When dealing with the resolution there is no possible downgrade in the quality of it in the real world view. We must also keep the safety in mind so that the user is not blinded if the HMD fails. During this time the user's eyes will be stimulated by the two cameras that will not align perfectly with the views seen by the naked eye. The video approach is more flexible in composing strategies, which is a serious problem in the optical approach. During this time the virtual images are not able to obscure the objects in the real world. Another problem is solved in the video approach and considered as a disadvantage in the optical approach that the optical system distorts the images when the user looks away from the center of the particular view. The further a person views, the larger the distortion he gets because the system has no digitized image of the real world in the system. In the video approach we must keep in mind that one delay is able to match with the other, so that the problem of the virtual image is solved by delaying the optical approach. Finally, the video that is able to capture the particular techniques using the digital camera in order to register with the real world strategies that are unavailable for the optical system.

Figure 6. Touch-screen interaction in public spaces

DIRECT PROJECTION SYSTEM

Augmentation can also be achieved by directly projecting graphics onto the real environment. Figure 4 shows an example of how the real world can be modified through controlled light that alters its appearance augmentations on different surfaces in the environment.

The advantage of this approach rises because direct projection eliminates the need for an optical or video-based combiner since the light is directly added to the real environment. The integration of real with virtual material takes place on the surfaces of the real environment and their combination is naturally perceived by the user. While the disadvantages occur in that projection depend on the availability of suitable projection surfaces, and on their geometry and appearance. It also relies on non-conflicting environmental lighting and the absence of objects in the optical path from the projector to the surface, to avoid occlusions (from becoming close) and shadows (Bandyopadhyay, 2001) and (Pinhanez, 2001).

SPATIALLY AWARE HANDHELD DISPLAYS

Handheld display is another approach used to the combination of computer graphics and the real environment tracking. It is used without the need of the optical or video see-through capabilities. Handheld display can provide useful context-sensitive information by leveraging proximity and relative position to real objects, despite of not matching Azuma's criteria for AR systems.

The first handheld prototype was presented in 1993, according to (Fitzmaurice, 1993). It was used to show how a tracked display could be moved over a physical paper map causing dynamically updated information to be presented about the different map areas as shown in Figure 8.5.

Figure 7. Touch light uses computer vision to track the user's gestures

The advantages of this approach came from the simplicity of tracking and calibration since the real and virtual are not visually registered with each other. The tracking resolution depends on the requirements of the application and how densely the real objects are positioned (Ishii et. al., 1997). However the disadvantage of this approach is the display occludes the real objects as the display is moved in the environment.

AUGMENTED REALITY INTERACTION TECHNIQUES

In general, interaction techniques are an important factor in any augmented reality application. It can improve the reality of these applications since the goal of augmented reality systems is to combine the interactive real world with an interactive computer-generated world in such a way that they appear as one environment. A number of different technologies that can enable interaction are described in the following sections.

Touch Interactions

Touch interaction is known as the easiest way of interacting with computer applications. According to Selker, (2008), this kind of interaction can directly be applied to graphical objects of interest without the need for special-purpose devices that users may need to operate in other forms of human-computer interaction. Improper calibration, parallax between the touch-sensitive layer and display element, low display resolution, low sensing resolution, and occlusion by the user's finger can prohibit accurate and precise interaction. All of these can be considered as the benefits of touch interactions (Figure 6).

For developing an AR application, (Seichter et. al., 2009) created a computer based vision that could track the library for the Augmented and Mixed Reality application which is known as SSTT also referred to as Simplified Spatial Target Tracker. This is known to run on a variety of devices that include cellular phones. It can even be optimized for speed in modular terms of the concepts that are used to track the targets. SSIT mainly focuses on robustness and is a lot faster than the other competitors in the market.

The Human Interface Technology Laboratory New Zealand HIT Lab NZ developed a commercializing technology that improves human computer interaction. The HIT LabNZ carries out the research with the help of the emerging technologies such as the Next Generation Video Conferencing, Immersive Visualization, and also the Human Robot Interactions. These Interaction Design methods are created in order to adapt the technologies to the benefit of the users so that the real world problems can be solved. The ultimate goal is to improve the individuals experience with the particular technology (Piumsomboon et. al., 2011).

Motion Interactions

The conceptual promise of using motion to interact with digital media is propel of the explosive adoption and growth of AR way forward. Companies empowered to offer more natural and unique virtual experiences for their audiences by launching their own versions of this type of AR through motion capture games and interaction technologies. Somehow, the use of motion capture technology seems to make sense as the next massive innovation waiting to explore leapfrog.com from the use of keyboard and mouse as data input devices to control personal computing devices. With processing power and webcam penetration seeing a steep rise, the creation of a more natural user interface (NUI) looks promising as another way to turn users' webcams into a data input device.

According to (Lucía et. al., 2011), motion interaction has been used and enhanced with other input devices to perform an AR system which allows the user to control the virtual objects movements, i.e. in the MoCap system the augmented mirror used to allow the audience to interacts with the avatar and the virtual objects. Audience also can talk with the avatar or walk around him. The use of motion interaction in some critical systems is used to simulate the direction of replacing things that are dangerous and require a very large cost which can eliminate important cues in performing tasks, such as the movement constraints, the response to a human interaction with the five senses i.e. hearing, smell, touch, taste and sight and respond to a given action (Yoki et. al., 2012). In addition, human eyes can be used in motion interaction as an eye tracker; it can recover the 3D positions of the user's eyes by illuminating the scene with multiple infrared light sources and capturing the reflecting light from the pupils on an image sensor that senses only infrared light (Yoki et. al., 2012).

According to Wilson (2004), human hands also are used in the motion interaction, as shown in Figure 7, using two infrared cameras and stereo computer vision to detect the position of the user's hands near the surface of a transparent display.

Handheld Devices

Handheld devices or (mobile devices) have huge benefits for AR since it consists of simple behavior that transmits button presses to complex input/output functionality and spatial awareness. Handheld devices can be used as a classical remote controls and advanced mobile devices.

In general, mobile AR is defined as any type of AR used on a mobile platform. It can be categorized as Cellular Phones, PDAs and Tablet PCs. All of these devices are considered as a replacement of the traditional, expensive and futuristic AR devices (e.g. Optical-HMD and Video-HMD), handheld AR displays are more accessible to the general public, and exhibit better learning curves than other AR devices (Wagner, 2005).

At present handheld and mobile devices are the best devices used in AR to develop games (Schmalstieg & Wagner, 2007). Early works rely on Head Mounted Devices (HMDs) for the visualization of the virtual elements while HMDs are still uncomfortable as the underlying technology is not mature enough (Cooper, 2004; Nilsen, 2005). Another drawback is the use of markers, as they are not elegant and constrain the design of the application. With technological and algorithmic improvements, it is now possible to use natural features instead of markers, and handheld devices for the visualization instead of HMDs (Wagner, 2008). For example in a AR game based application, most mobile AR games require the user to look at the user's hands through the handheld camera, and there are elements to move or interact with that are part of the game.

Figure 8. Black and white markers

Figure 9. Assembly instructions application based on marker based augmented reality

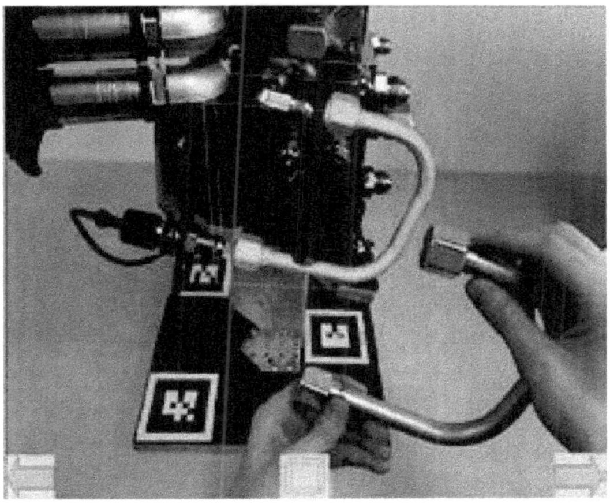

Furthermore, the mobile AR is used in navigational systems to help in bringing the user somewhere and typically work in one of two ways. The simpler way is via a compass. When the user lifts the mobile device to look at an area and enters an address, a compass will appear on the screen and show the user where to walk or drive (Wagner, 2008). Another navigational mobile AR system shows a line on the mobile device's screen telling the user where to go, similar to street lines from a GPS device (Jang, 2012).

Figure 10. The process of detecting the markers by the computer system

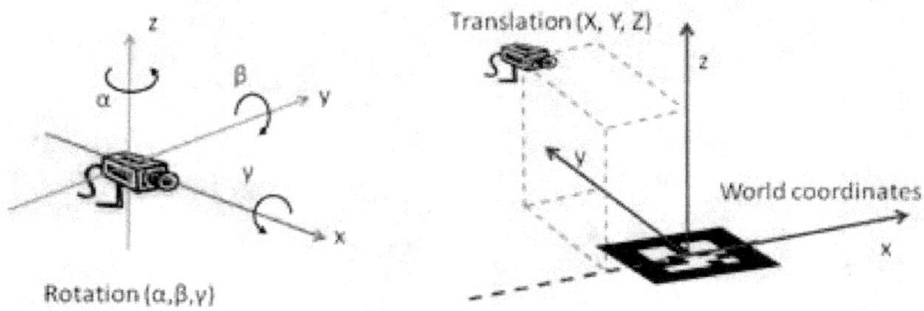

Speech Input

Speech interfaces in general are employed as a command-based interaction style, where the user executes actions through verbal commands. Interactivity can, however, also be achieved through analysis of non-verbal features. Most of researches done in speech interaction focused on how speech and gesture input can be used to manipulate virtual objects. One of the first interfaces to combine speech and gesture recognition was the media room. The media room allowed the user to sit inside the computer interface and interact with the computer through voice, gesture and gaze. "Talk and Draw" application allows users to draw with a mouse and use speech input to change interface modes (Salisbury, 1990). Similarly QuickSet application that combined speech and pen input for drawing on maps in command and control applications (Cohen, 1997).

However, there is a near limitless amount of different ways a user could interact with an AR system. Normal ways of human-computer interaction, like WIMP (Windows, Icons, Menus, and Pointers) can be supplemented with anything that user's senses can take as input and with any action a user could come up with. One of the goals of AR is to come up with interaction that feels as natural as possible to the user, which is often for the most complex part (deLandgraaf, 2003/2004). This is one of the considerations of the research done in this research.

Figure 11. Example of Marker-less AR (MAR) (Dolz, 2010)

Figure 12. Augmented reality geo-location application working under android (Dolz, 2012)

IMPLEMENTATIONS OF AUGMENTED REALITY APPLICATIONS

Where image recognition is used to align a virtual object (optical tracking), there is a distinction between systems that identify an artificial fiduciary marker (2D matrix code or Light Emitting Diode) placed at locations or on objects in the real world and those which use natural feature detection to identify unaltered real world objects such as book covers, posters or landmarks that have no artificial makers to assist object recognition (Butchart, 2011). In each case, the 2D or 3D virtual object appears to the marker or natural feature when seen through the AR viewport. In the following sections, the two main approaches used in AR to track objects are discussed.

Marker-Based Augmented Reality

AR presents information in a correct real world context. To achieve this task, the AR system needs to know where the user is located in the environment, as well as what the user is looking at (Höllerer & Feiner, 2004). In general, the user explores the environment through a display that portrays the image of the camera together with augmented information. For this, the system needs at first to determine the location and orientation of the camera with a calibrated camera, so the system then will be able to render virtual objects in the correct place. A number of different tracking methods have been introduced. These methods are classified based on the equipment used in sensor tracking, visual tracking and hybrid methods. Since in most AR setups, the camera is already part of the system (Erkan et al., 2010), the visual tracking methods are of special interest and most used in AR applications.

As an example the marker (Figure 8) (which is considered as an easily detectable predefined sign in the environment) is used in AR system for tracking objects (Figure 9). The computer system can detect these markers from a video image using image processing, pattern recognition and computer vision techniques (Figure 10). Once detected, it then defines both the correct scale and pose of the camera.

Another approach in visual tracking is the feature-based or (model-based methods). In this approach, the system has a model of the scene or part of the scene. It compares visual observations with the model and finds the best match, which then defines the pose (Petersen, 2008).

The popularity of marker-based systems is also partly explained by the fact that they are good and easy to implement. Well-known marker-based toolkits are available with a variety of libraries such as AR Tool Kit, ALVAR and ARTag.

Marker-Less Augmented Reality

In order to perform the object tracking, marker-less AR systems rely in natural features instead of markers defined previously. Therefore, there are no ambient intrusive markers which are not really part of the environment. Marker-less AR counts on specialized and robust trackers already available.

The advantage of the marker-less systems (Figure 11) is the possibility of extracting from the environment characteristics and information that may later be used by them (Dolz, 2012). However, among the disadvantages, that can be considered for marker-less AR systems is that tracking and registration techniques become more complex than in marker based system.

The idea of marker-less AR is very straight forward. It relies in natural features of the image or object to be tracked, like the edges, corners, or textures. This approach of tracking has been used in developing many AR applications. AR Browser SDK can add AR Geo-location view to Android or iOS application in less than 5 minutes (Figure 12). It is a fully customized framework with easy to use API (Application Programming Interface). The framework takes care of all the complex functionalities of the AR browser (Simon, 2000). In an example application, 'planar local image patches' are tracked by similar application for Region Tracker and then brought to complete correspondence after that it is possible to add virtual texture to them (Ferrari, 2001).

AUGMENTED REALITY SOFTWARE AND TECHNOLOGIES

The measure of the AR system has to do with how it is able to cope with the changes in the real world. This software that is able to derive the real world coordinates which are independent from the camera. This process can be referred to as registration, which utilizes numerous methods of computer technology that are closely related to video tracking (Azuma, 2001). Many technologies were used to integrate augmentations with the real world, such as (ARToolKit, FLARToolkit, FLARManager, Papervision3D, Daqri, Unity3D, Wikitude, ARcore, Vuforia, Maxst 3D, DeepAR, EasyAR, Xzimg, Adobe Flex, Adobe Flash, Action scripts and Adobe Flash-Builder 4.

AUGMENTED REALITY AND EDUCATION

Despite the presence of a number of educational institutions that still rely on old-fashioned\traditional teaching methods as lecturing process, taking notes, reading books and taking final exams, nowadays, there are many educational institutions competing in implementing modern scientific techniques and taking the effort to engage their students in meaningful activities that could reach students at multiple levels of education.

It is a general fact that in terms of teaching students, the more senses are involved in the learning process (sound, sight, touch and emotions), the more effective the learning experience would be. According to Manfred, (2006), AR can be integrated as a teaching platform into the instructors and students environment for displaying audiovisual and multimedia content in line with the needs in the relevant areas. AR interfaces enhance the real world experience, unlike other computer interfaces that sometimes draw users away from the real world and onto the screen. According to Mariano et al., (2006), AR appears as an emerging futuristic technology that promises to make "educational immersion" available to practically everyone.

A lot of development has been made on AR since its inception. One can easily find many examples of AR applications in games, advertisement and entertainment. Award for Best Developer Tool or Platform in 2012 has been given to Wikitude Architect, while Metaio Mobile SDK + junaio remain first runner-up and Layer & Layer Creator was second runner-up. Award for Best Augmented Reality Application (excluding Browsers) in 2012 has been given to Urban Troops (Future Reality Games), while Starwalk (Vito Technology Inc.) remain first runner-up and Bugs Mayhem (AR23D) was second runner-up. Award for Best Example of Augmented Reality in Marketing in 2012 has been given to Ikea Interactive Catalogue (McCann Erickson), while King of the jungle (National Geographic) remains first runner-up and Angry Birds (Nokia) was second runner-up (Lester, 2012).

In the field of education there some applications which are specifically designed for educational purpose and some other which could be used to enhance learning experience. Occupational Safety Scaffolding, School in the Park (Augmented Reality Experience), Mentira, Driver's ed, Handheld Augmented Reality Project and Project Glass are few of many that helps the learning process (TeachThought, 2012).

Since 1990 many VR/AR systems dedicated for educational use have been introduced (Bell & Fogler, 1995), (Byrne, 1996), (Loftin et. al., 1993), (Winn, 1997), (Mantovani, 2001), (Billinghurst, 2002), (Liarokapis, 2002), (Kaufmann, 2003), (O'Brien, 2005), (Dünser & Hornecker, 2007), (Squire & Jan, 2008), (Oh, 2008) and (Maqableh, 2012).

Some good examples of VR systems include Narrative-based, Immersive, Collaborative Environment for teaching Biology which is considered as an immersive system in which children can plant flowers, move clouds to water them or move the sun to make them grow faster. It was developed by Roussos et al., (1997).

CyberMath for mathematics education presents an avatar-based shared virtual environment for improving mathematics education. It was developed by Taxén & Naeve in 2001. In 2002 Studiers tube was introduced as a system that presents several capabilities for AR collaborative interaction. Kaufmann (2003) developed an AR system "Construct3D" for geometry education.

BENEFITS OF AUGMENTED REALITY

The use of new technologies in education can possibly enrich the modern educational theories and make the learning process to be more active, attractive, motivating, simulating, and meaningful to the student. Although it is usual in education to be a bit cautious when new kinds of technology are accepted, we tend to hide the truth that it can change the educational system from the memory based learning to more motivated and creative education (Chen, 2006).

The technology that is developed must be transparent and should not encompass education where there is emphasis present on the teaching material. The learners must not spend their time getting used to the new methods of learning and must strive to develop a base that deals with the already established learning tenants (theory). "*If the student is overly conscious of the technology permeating the learning experience, this can inevitably lead to dissatisfaction*" (Ramirez, 2011).

An AR as an advanced technology that allow the users to interact with virtual and real world in real time application (Azuma, 2004). It can bring more natural experience, raises attention and motivation to students with a high potential to enhance the learning experience (O'Brien, 2005). AR has the potential to engage, stimulate and motivate students to explore class materials from different angles (Kerawalla, 2006). Shelton, (2002) reported that AR technology could also be used to help in teaching subjects where students could not feasibly gain real-world first-hand experience (e.g. astronomy and geography). On the other hand for the effectiveness and attractiveness, AR supports students in learning in a real life (Dünser, 2007).

According to Yelva et. al., (2011), when the learners actively involved in an experience in the AR environment, they must remember to retain all the information that is given to them. It may be that the abstract ideas could be a bit difficult to understand, but they can be presented in an enhanced learning environment that provides access to the historical artifacts and the invisible data (Billinghurst, 2004). In addition as Eike & Fotis (2010) reported that the level of understanding of the students in complex learning issues could be increased through an effective use of engaging audio-visual augmentation, which should be "*attractive and highly motivating to today's generation of media-conscious students*".

The potential benefits of AR applied to education can be summarized in the following points (Ramirez, 2011):

1. **AR can improve the multi-modal** (provides the user multiple modes of interfacing and interaction with a system) visualization of difficult theoretical concepts. Since the multi-modal visualization allows students to switch interactively between different visualization media.
2. **Tangible (physical existence)** AR interfaces can help students to explore the multi-dimensional augmentation of the teaching material in various levels of detail, so this could transfer the theory into tangible examples.
3. **Natural interaction with multimedia** representations of teaching material allow the students to navigate through the augmented lecture data and therefore concentrate and study in detail of any part of the lecture.
4. **Effective collaboration** and discussion amongst the participants. Even though the growth of AR technology has been studied for more than forty years, but the focus of the formal evaluation on AR application has only recently developed. In the following section the use of AR in engineering education is discussed.

AUGMENTED REALITY IN ENGINEERING EDUCATION

AR has more potential in enhancing a person's perception of virtual and real surrounding and immediate environment, than Virtual Reality (VR) that replaces the perception of the world with an artificial one (Potter, 2011). In addition AR support new type of interfaces i.e. tangible interface which is able to display relevant information at the appropriate time and location, offers many potential applications, which can be used to aid engineering education, training, repair or maintenance and manufacturing.

As shown in previous research works, AR is a useful technology for improving educational and training systems, and it has been introduced as educational application of visualization techniques. For example, Klopfer & Jenkins, (2002) studied the potential of AR in environmental engineering education where students used location-aware handheld computers (equipped with GPS) to physically move throughout a real world location. Kaufmann, (2003) developed Construct3D, a mobile collaborative AR application designed for mathematics and geometry education. Liarokapis et al., (2004) presented an educational application that allowed users to interact with 3D Web content in VR and AR. Four mechanical engineering themes (machines, vehicles, platonic solids and tools) were also illustrated to demonstrate the use of the system to support learning. Hughes et al., (2005) described their efforts to transform core mixed reality technology and methods into the diverse applications of urban terrain, used for military training and situational awareness, and community learning. Chen, (2006) investigated how students interact with AR and physical models and evaluated their perceptions in learning about amino acids. The results showed that some students liked to manipulate AR by rotating the markers to see different orientations of the virtual objects and some preferred to interact with physical models in order to get a feeling of physical contact.

Valentini, (2010) discuss a methodology to enhance multibody systems simulations using AR implementation, the results showed that AR can be useful for didactics purposes to support the explanation of motion of complex mechanisms. Jorge and Manuel, (2011) have introduced and augmented book with its matching AR application for knowing and learning standardization regarding mechanical elements. The results revealed that students showed more participation and motivation than in the teacher's theoretical classes. Henderson and Feiner, (2011) describe an AR application for maintaining and repairing a portion of a military vehicle, the evaluation showed that the AR application allowed mechanics to locate individual tasks in a maintenance sequence more quickly than when using an improved version of currently employed methods.

For the engineering education, instructors notice that new generations of students in engineering subjects show more interest and pay more attention when they use CAD tools, multimedia material and virtual platforms. However the students show apathy (lack of interest) and less motivation when using traditional learning tools (Jorge & Manuel, 2011).

The use of AR technology in engineering education allows the students to have the establishment of contact and to adhesion with the didactic material (resources that can help in developing understanding) while studying tasks (Jorge, 2011). Jorge et al., (2010) reported that, AR technology is an attractive and motivational technology that facilitates the learning of the basic contents of engineering subjects, it can help to improve students' spatial abilities and it allows students to solve and visualize engineering problems in an autonomous and active way.

AR technology has been used to develop many other engineering applications, for example Margot, (2001) develop a tangible media for engineering drawing lessons that allow engineering students to design mechanical devices that support interaction and user inquiry.

Liarokapis, (2002) presented an experimental system that combines the use of AR technologies with XML metadata. The integrated system provides the content for multimedia presentation in a virtual environment to help the students to visualize basic and advanced principles and therefore test their learning throughout virtual scenarios.

Additionally Liarokapis et al., (2004) have examined the use of AR as a tool for engineering where Web3D and AR technologies allow students to explore mechanical engineering concepts. Martin-Gutierrez et al., (2010) implemented an AR application called "AR_Dehaes". The application was developed based on performing graphic engineering exercises using AR to improve the spatial skills of students in understanding the contents of geometric graphics. Heen, et al., (2011) designed an AR based engineering graphics system whereby the system discusses the potential and challenges of using AR in engineering graphics education and how it can be used to improve the spatial awareness and interest of learning.

According to Wolfram, (2002) AR can provide a unique insight into the way the world works and the simulation of complex and dynamic systems become possible. Augmented reality simulation can provide: domain-related practices, ownership of the inquiry, coaching and modeling of thinking skills, an opportunity for reflection, open-ended dilemmas and motivation for learning engineering context (Kurt and Eric, 2004).

INTERFACE DESIGN FOR LEARNING ENVIRONMENTS

Manjit, (2007) stated that the interface is the method that the learner has to incorporate for utility of the software package/tool for a learning environment; no matter it is usual CAL or other forms of computer based learning aids.

The genuine functional performance not the exact implementation of the software package/tool cannot be entirely separated from the user interface (Kinshuk, 1996). 'Leaks Through' of the functionality into the interface is also possible. Kinshuk, (1996) noted that what is heard, done and seen by

Figure 13. A rigid body in a plane has three DOF

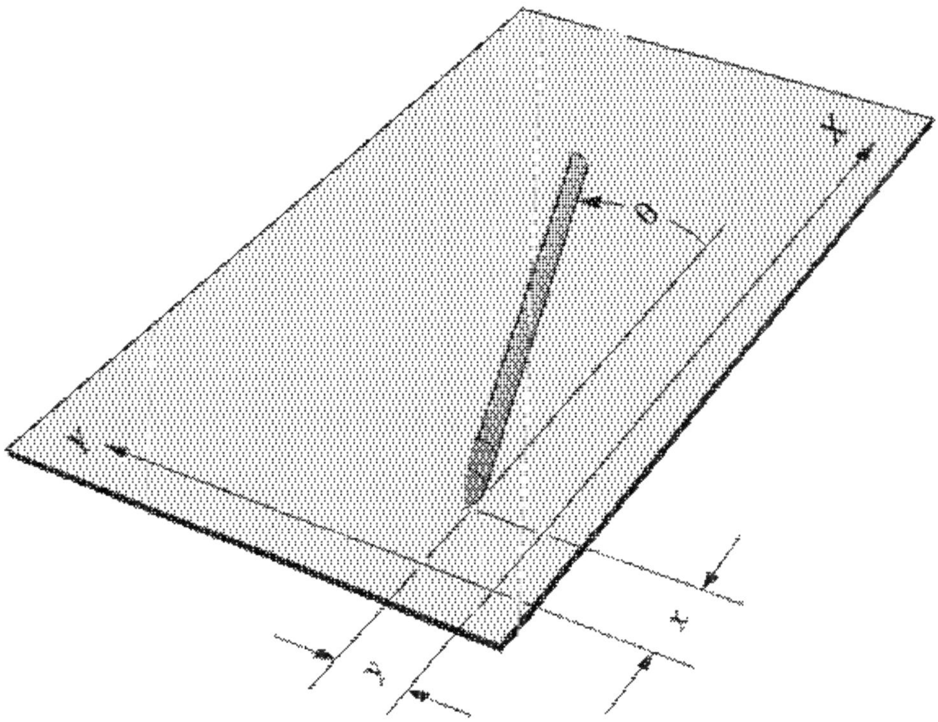

the learner-as-user via the interface resultantly shapes the progress of the learner's model of the system and the domain being taught.

On the other hand O'Malley, (1990) argued that vigilant consideration must be paid to three types of representations at the interface (Representations of the semantics of the domain to be learned, Representation of the system's functionality and representation concerns what the user is expected to do by interacting with the system).

Matcha, (2012) described that; AR interfaces contain a rich and complex type of object and interaction. The concept of co-existence of virtual information and physical objects of AR enhances the learning environment. Furthermore the potential of AR to support collaborative learning could be done by allowing the learner to share their knowledge and skills to achieve or solve the problem together. Other user interface practitioners also agreed that the TUI's and AR interfaces can improve the collaborative learning characteristics including engineering students Makina and Sazilah, (2011). As such the TUI approach was chosen for designing the ATAPS software.

The interface of an AR system allows the user to modify their own reality, extending it in directions never imagined before Griffin (2013). Graphical user interfaces allow us to visualize complex data sets, in new ways. Griffin (2013) further stated that one achieves the skill to understand physical environment and one's place within in it, when correspondence of data set and immediate physical environment happens. The use of symbolic languages and well-designed semiotic icons permit learners to correspond without regional linguistic variation, achieving accuracy of expression and clearness in the transfer of meaning that is merely not viable in the "natural" world (Griffin, 2013).

THE FUNDAMENTALS OF LINKAGE MECHANISMS

According to Manjit (2007): "When tutoring a student about solving an engineering problem, a human tutor is expected to gauge the student's background knowledge, deliver relevant course material at the correct level of detail, and clarify student's misunderstandings".

The very first tools that were created by humans were rocks, bones or sticks; the first machine would be a stick working on principles of lever and fulcrum and the first invention was wheel. However control systems are now planted in advanced machines which were initially mechanical and this enables machines to work without the help of humans in their all tasks. The number of degrees of freedom (DOF) could classify the types of mechanical systems. Whereas, DOF is the number of independent parameters (measurements) that are required to uniquely define its positioning space at any instant of time (Norton, 2009). Figure 13 shows a rectangle on a flat piece of paper with 2-D coordinate system.

Norton, (2009) stated that if pencil is constrained to always remain in the plane of the paper, three parameters (DOF) are needed to entirely define the position of the rectangle on the paper, two linear coordinates (x, y) to define the position of any one point on the pencil and one angular coordinate (θ) to define the angle of the rectangle with respect to the axes. The minimum number of measurements needed to define its position is shown in the figure as x, y, and θ. This system of the square in a plane then has three DOF (Norton, 2009).

Figure 14. Four bar linkage rigid bars

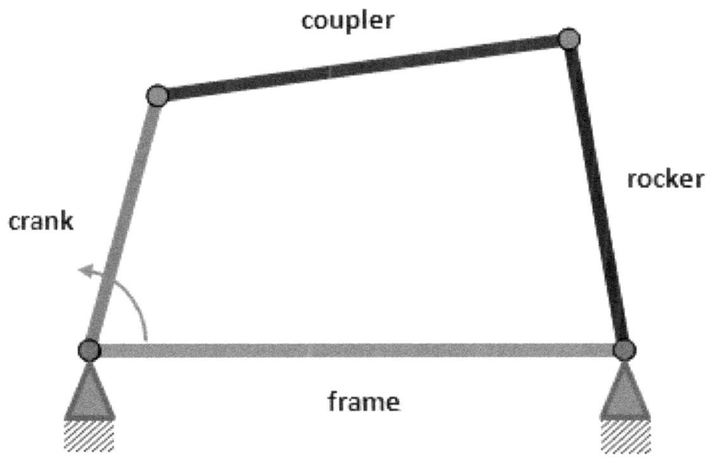

Figure 15. Four bar linkage configuration

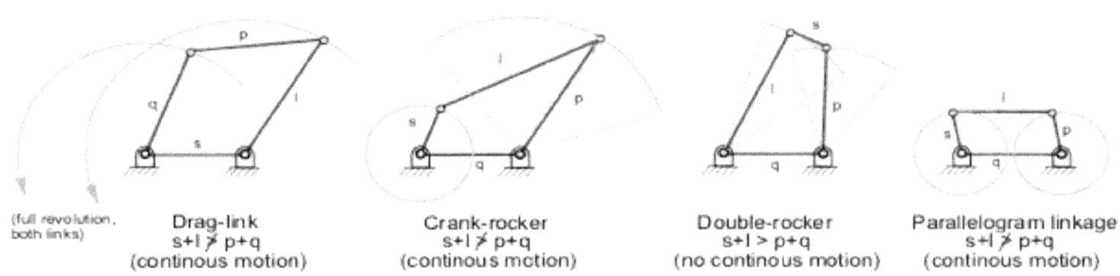

In this example the rectangle is a rigid body, or link, which for purposes of kinematic analysis would be presumed incapable of deformation. This is only a convenient fiction to allow user to more easily define the gross motions of the body. Later on any deformations due to external or inertial loads onto our kinematic motions to acquire a more absolute and correct picture of the body's behavior could be superposed (Norton, 2009).

Types of Motion

In the general case, a rigid body free to move within a reference frame has multifaceted motion, which is a synchronized amalgamation of rotation and translation. There could be rotation about any axis (any skew axis or one of the three principal axes) in three-dimensional space, and also simultaneous translation that may be resolved into components along three axes. In a plane, or two-dimensional space, complex motion becomes a combination of simultaneous rotation about one axis (perpendicular to the plane) and also translation resolved into components along two axes in the plane (Norton, 2009). According to Robert, (2003) the motions and rotations of a rigid body have three types:

1. **Pure Rotation:** The body possesses one point (center of rotation) that has no motion with respect to the "stationary" frame of reference. All other points on the body describe arcs about that center. A reference line, drawn on the body through the center changes only its angular orientation.
2. **Pure Translation**: All points on the body describe parallel (curvilinear or rectilinear) paths. A reference line drawn on the body changes its linear position but does not change its angular orientation.
3. **Complex Motion**: A simultaneous combination of rotation and translation. Any reference line drawn on the body will change both its linear position and its angular orientation. Points on the body will travel nonparallel paths, and there will be, at every instant, a center of rotation, which will continuously change location.

Translation and rotation represent independent motions of the body. Each can exist without the other. If we define a 2-D coordinate system as shown in (Figure 14), the X and Y terms represent the translation components of motion, and the θ term represents the rotation component (Marghitu, 2009.

Four-Bar Linkage Mechanism

A four-bar linkage (4BL) is the simplest movable closed chain linkage. It consists of four bodies, called Links (the most common links are known as binary, ternary, quaternary and pentanary links) or (Bars), connected in a loop by four Joints (Joints also called nodes; it's the pivot appear at the end of each Link). Generally, the joints are configured so the links move in parallel planes, and the assembly is called a planar four-bar linkage and if the linkage has four hinged joints with axes angled to intersect in a single point, and then the links move on concentric spheres and the assembly is called a spherical four-bar linkage.

The 4BL linkages are important mechanisms found in machines and are important concept in mechanical engineering, and other disciplines that make use of connected rigid objects. It consists of four rigid bars connected by pins such that the bars can rotate, potentially 360 degrees, in the same plane. As shown in Figure 15 they are constructed from four links connected in a loop by four one degree of freedom joints. One link, called the frame, is fixed. The crank (also called the input link), is connected to the frame and is the driving bar. The rocker (or follower link) is pulled by the crank via the coupler, which connects the crank and rocker. The reason for the terminology is because the configuration (Figure 16) can lead to different modes of operation as the crank is rotated, depending on the relative lengths of the bars (here s=shortest link, l=longest link). In particular, the crank-rocker mode can be used to generate a back-and-forth rocking motion from continuous rotation. More information on the mechanisms can be found in well-known engineering text books.

DEVELOPMENT OF 3D MODELING AND ANALYSIS SOFTWARE TOOL

This section describes the approach taken for using the computer-based technologies in engineering problem solving. While many software packages have been developed and used for the purpose of student learning in engineering courses, this approach came to enhance the student's learning ability and to improve their visualization and imagination of the mechanical concepts.

The ATAPS package was developed based on the use of the augmented reality technology integrated with FLashBulider4.6, OpenGl and Action Scripts. This combination was chosen in order to build an Interactive application and gives the ability to students to interact with and visualize the 3D-simulation in real time.

As stated before in the section titled "Problem Statement", the Chapter highlights some of the difficulties that the engineering students face during learning mechanics subject. Dünser et. al., (2007) stated that affordances through tangible user interface (TUI) has the advantage to be used in 3-Dimensional learning and construction environments, this research also aims and suppports in the benefits that can be obtained from the use of 3D-simulation and modeling in the real time environments to reduce or minimize the difficulties faced by students.

In the design of the ATAPS package as an engineering problem solver, several key techniques have to be considered as following:

1. Real-time tracking and modeling: computer can render and register a virtual object in the real environment quickly and properly.
2. Interactive technique: user can interact (touch to print) with the system components, and changing the mathematically analyze the selected problem in real time and print the outputs on properties of the virtual object smoothly.
3. Real-time 3D simulation technique: system can generate 3D models of the selected problem whenever the properties changed.
4. Real-time analysis: system can analysis the problem equations and provide the analysis results in real time environment.

Description of the Integrated Technologies

The ATAPS package in this research was developed in integration with some other application and open source libraries such as FlarManager 6.0 to build the augmented reality environments, FLashBulider4.6 as the main interface environments, OpenGL used for the mechanical structure simulations and Action scripts used in writing the program codes.

FLashBuilder 4.6

Adobe Flash is a multimedia platform developed and distributed by Adobe Systems. Since 1996, to add animation and interactivity to many computer applications, Flash has been a famous method. Usually web pages, animations and advertisements have been created by the use of Flash. Recently, to integrate videos into web and to create rich Internet applications, Flash components have been utilized.

Flash has been a popular software since its creation. Several software and systems are able to create or display Flash content, in addition to that the Adobe Flash Player, is freely available for most common web browsers, mobile phones and for other electronic devices.

For the programming and the applications development the Adobe Flex software development kit (SDK) was used because the conventional application programmers found it difficult to utilize the animation metaphor on the basis which the Flash Platform was initially designed. Flex seeks to lessen this difficulty by giving a workflow and programming model that is recognizable to these developers, in addition the Flex applications can be written using Adobe Flex Builder or by using the freely available Flex compiler from Adobe.

The Flex SDK is available with a set of user interface components that contains several text controls, buttons, trees, data grids, list boxes, and a range of layout containers. Graphs and Charts and offered as an add-on.

Action-Scripts

Based on ECMA Script, Action-script is a scripting language. Action-script is used mainly for the creation of websites and development of software employing the Adobe Flash Player platform (in SWF files implanted into Web pages), but is also employed in few database applications (like Alpha Five), and in basic robotics (like the Make Controller Kit). Actually developed by Macromedia, the language is now in possession of Adobe (which gets hold of Macromedia in 2005).

For controlling simple 2D vector animations, Action-Script was originally designed by Adobe Flash (previously Macromedia Flash). Later on more functionality was added that allows the development of Web-based games and rich Internet applications with streaming media (such as video and audio).

OpenGl

Open Graphics Library known as OpenGL is a specific standardization defining a cross-language, cross-platform API for developing applications that generate 2D and 3D computer graphics. It is extensively used in CAD, information visualization, Virtual Reality, scientific visualization, flight simulation and also in video games that defends with Direct3D on Microsoft Windows platforms.

Away3D

Developed for the Adobe Flash platform in ActionScripts 3, Away3D is an open source 3D graphics engine. It runs in recent web browsers that utilize Adobe Flash Player. Away3D can execute various 3D computations and it can also be employed to make 3D models.

Figure 16. (AR ATAPS package) user interface in real time environment

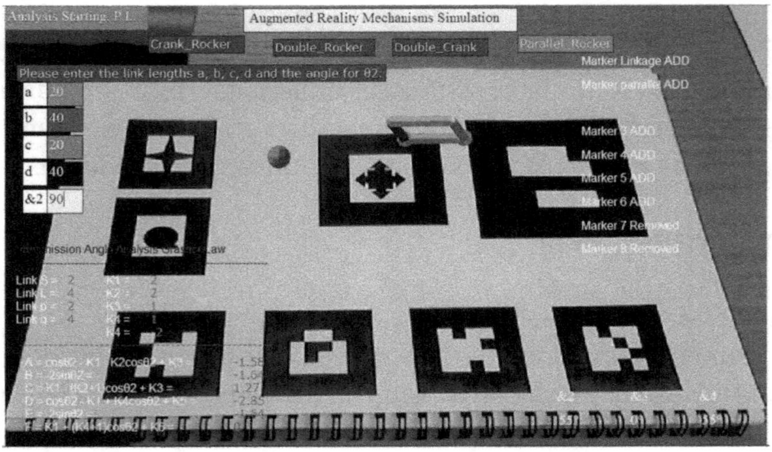

Papervision 3D Code Library

Papervision 3D has complicated geometrical equations that basically enlarges, shrinks and warps objects in a scene which in turn makes them appear to be moving in a 3D space. Besides being a well-crafted set of instructions to pull off the expanding, shrinking and warping effect of objects to make a scene appear 3D, it is also packed with classes and methods to help a user to pull off 3D effects.

ARToolKit and FlarManager

ARToolKit is a free, open source C software library that uses computer vision techniques to calculate camera pose (position and orientation) relative to a black square marker. It is fast enough to perform this calculation at 30 frames per second on a normal desktop. ARToolKit is typically used for applications that augment video of the real world with computer generated objects. FLARToolkit is the newest in series of ports of ARToolkit.

FlarManager is a lightweight structure that enables creating FLARToolkit augmented reality applications simpler. Its purpose is to offer faster access to the technology, to permit a focus on new forms of interface through marker detection and augmented reality.

Figure 17. The user interface components in offline mode

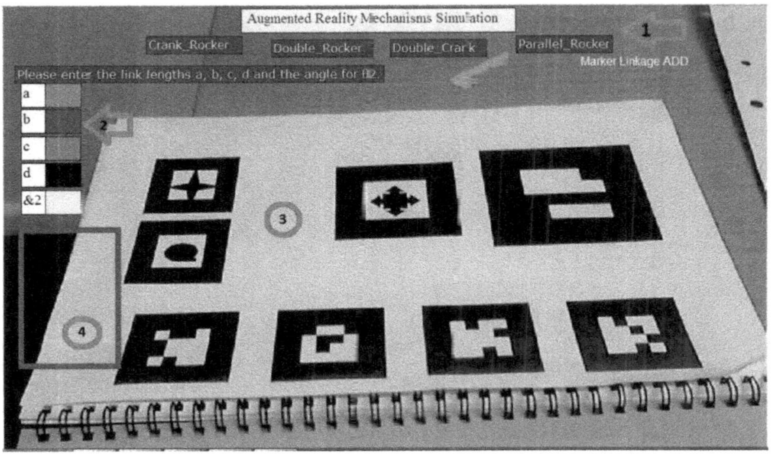

Web Camera

A web camera should be installed for developing any augmented reality applications. It's used to allow the users to see the surrounded environment, and to capturers the image of the markers placed in the real time environment.

Setup the User Interface

The user interface as mentioned by (Guralnick, 2006), is a crucial part of a user's experience with any piece of software. The interface design and learning effectiveness are potentially correlated with each other. This in result makes the interface design critically important for educational purposes.

One of the important design approaches is the Tangible User Interfaces (TUI); that allow the user to interact with the computer through physical objects rather than mouse and keyboard. Billinghurst, (2001) states that the Tangible Augmented Reality is a combination of an Augmented Reality System and a Tangible User Interface, where the users interact with the virtual objects by manipulating real objects. The proposed user interface in this research is developed based on the TUI approach enhanced with the touch to print interaction style. The following points summarize the features of the proposed user interface:

Figure 18. The user interface components in online mode

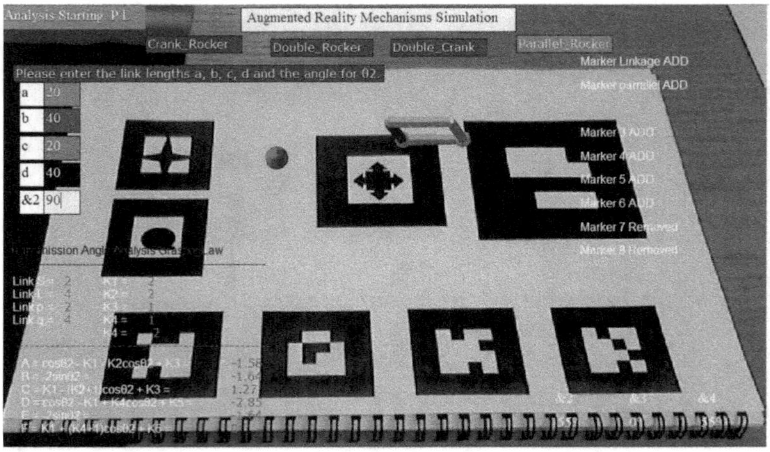

The Use of Real Time Environments

It was notified in many researches that the education sector is confronted with new and powerful technological tools that could be used as additional teaching aids for learning purposes. The use of multimedia and virtual reality (VR) that was founded to be useful in learning engineering (Huber & Dietmaier, 2010), (Manjit, 2009).

This research aims to discover the effectiveness of using augmented reality technology as a new teaching aid for learning purposes due to the useful properties of this technology in the educational area that been briefly discussed in detail in the chapter.

One of the main characteristics that this research seeks to employ in the development of the user interface is the ability of the augmented reality environment to view and interact with the three-dimensional virtual elements in real time environment. Figure 16 shows the user interface designed for the AR ATAPS package.

The User Interface Design Approaches

The instructor teaching the Mechanics Dynamics subject at UNITEN found that the students have difficulty in understanding and solving the problem of the four bar linkage (4BL).

In general the students are required to study three main concepts of the 4BL theory i.e. different link and joint types which form the 4BL mechanisms, the kinematics and kinetics of closed planar mechanisms and the static and dynamic force analysis of closed planar mechanisms.

The understanding of the 4BL theory will be achieved only if the students were able to:

1. They are able to clearly identify the different link and Joints
2. They are able to recognize the kinematics and kinetics of the 4Bar mechanisms.
3. They are able to analyze the static and dynamic force of the 4Bar mechanisms.

In order to achieve the previous goal, the user interface was designed as a combination of two approaches of user interface design approaches i.e. (Graphical User Interface and Tangible User Interface).

The Graphical User Interface (GUI) has been used to implement the screen components since it has been accepted that GUI can provide natural and easy means for users to communicate with computers (Kinshuk, 2003). In another study conducted by Manjit, (2004) GUI was also founded to be suitable for engineering students to interact with.

The Tangible User Interface (TUI) was used for the interaction purposes with the AR ATAPS package in this research, since it can offer a variety of interactions. Xu (2005) in her research found that TUI allows users to solve problems with concrete physical objects and physical action when they fail using more abstract representations and complex syntax. Therefore TUI can empower users with the control of the computing environment as they will feel and own the environment and will be actively engaged and not losing their interest easily.

Table 1. Set of data collected from the survey regarding question 1

Q No.	Text Book (1)	Current Packages (2)	AR ATAPS package (3)	Total
1	**The way the problem is presented by the (1, 2 and 3) makes it easy to work out the solution.**			
	4.033	3.300	1.867	
2	**The (1, 2 and 3) used during the class time/Lab; make it easy to understand the engineering problems.**			
	3.667	3.267	2.267	
3	**Most students would find it easy to solve Engineering Mechanics problems using (1, 2 and 3)**			
	3.700	3.333	2.667	
4	**After using (1, 2 and 3), I now understand how to solve Engineering Mechanics problems.**			
	3.667	2.967	2.567	
5	**I can easily understand the concepts of the four bar Linkage by using the (1, 2 and 3)**			
	3.500	3.367	2.100	
Total	**18.567**	**16.233**	**11.467**	**46.267**
Mean	**3.713**	**3.247**	**2.293**	

Table 2. Question 1, variance table

Source Of Variation	Sum Of Squares	Degree Of Freedom		Mean Square		F Ratio
Between Tools SS	5.238	k-1 = 3-1	2	s_1^2	2.619	s_1^2/ s^2
Within Tools SS	0.690	k(n-1) = 3(5-1)	12	s^2	0.058	45.536
Total SS	5.929	kn-1	15			

Constructing the User Interface

The AR ATAPS package interface has been developed using Flash-Builder 4.6 with Action-script3.0 codes. It was deigned to be user-friendly by providing the user all of information needed to solve the selected engineering problem.

The user interface has two modes offline mode and online mode, has been divided these situations into four main sections that are similar in appearance and differ in the nature of the information provided. (Figure 17) shows the Offline mode while (Figure 18) shows the online mode.

Figure 19. Research question 1

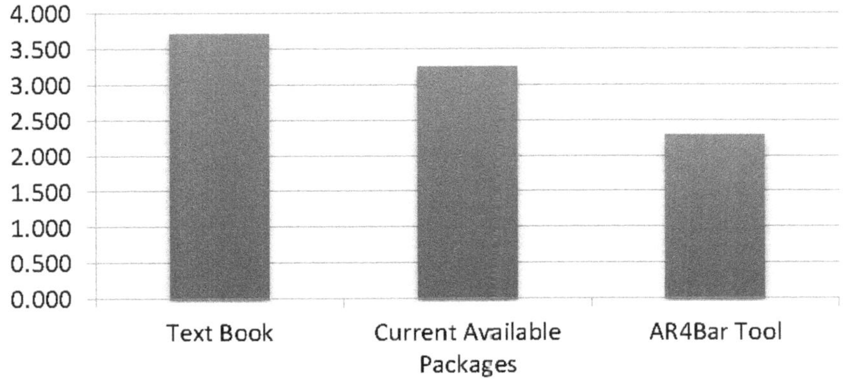

1. **Mechanisms Section**: It shows the different types of the four bar linkages i.e. (Crank Rocker, Double Rocker, Double Crank and Parallel Rocker).
2. **Data Section:** It shows the data entry section, used to pass the equation parameters into the program.
3. **Modeling Section:** This section represent the Real time environment where the different types of the mechanisms will be simulated and modeled.
4. **Analysis Section:** This section holds the real time data analysis of the different types of the mechanisms.

1. **Mechanisms Section**: Shows the current type of the 4BL matched with the parameters in the data section. (Red Square)
2. **Data Section:** Shows the parameters entered by the user.
3. **Modeling Section:** The Real time environment shows the 3D-Virtual models of the 4BL. And it also shows a red sphere indicating that the simulation process of the 3D-Model is started.
4. **Analysis Section:** Shows the detailed analysis of 4BL equation.

Table 3. Set of data collected from the survey regarding question 2

Q No.	Text Book (1)	Current Packages (2)	AR ATAPS package (3)	Total
1	I can easily identify the different types of the four bar mechanism using (1, 2 and 3)			
	3.700	3.167	1.867	
2	I can easily visualize the four bar linkage from (1, 2 and 3)			
	3.667	3.367	2.333	
3	I can easily identify the elements of the four bar Linkage in the (1, 2 and 3)			
	3.200	3.233	2.667	
4	I can easily imagine the Kinematics of the four bar Linkage using the (1, 2 and 3)			
	3.933	3.733	2.033	
5	(1, 2 and 3) is interactive and use multimedia.			
	3.967	2.900	2.767	
6	(1, 2 and 3) supports the (2D animations) in its contents.			
	4.567	2.500	2.633	
7	(1, 2 and 3) supports the (3D animations) in its contents.			
	4.533	3.900	2.133	
8	I can Analyze the dynamic problems using (1, 2 and 3)			
	3.667	2.967	2.567	
Total	**31.233**	**25.767**	**19.000**	76.000
Mean	**3.904**	**3.221**	**2.375**	

DATA ANALYSIS AND DISCUSSION

Data analysis is divided into six sections according to the main parts of the questionnaire namely Experience with Computing, Learning Experience during Mechanical Engineering Courses, Computing Knowledge and AR ATAPS package, Research Question 1, Research Question 2 and Research Question 3 as stated below. The first three sections were analyzed by simple percentage analysis along with graphical representation and last three sections were analyzed by ANOVA.

Table 4. Question 2, variance table

Source Of Variation	Sum Of Squares	Degree Of Freedom		Mean Square		F Ratio
Between Tools SS	9.389	k-1 = 3-1	2	s_1^2	9.389	s_1^2/s^2
Within Tools SS	3.702	k(n-1) = 3(8-1)	21	s^2	3.702	26.625
Total SS	13.091	kn-1	23			

The research questions addressed in this research are as follows:

1. Can AR ATAPS package help learners in visualizing and understanding the selected difficult engineering problems of the subject matter?
2. Do virtual problem solving aids with mixed realities and touch to print tangible user interface enhance student learning in engineering?
3. Do the learners accept the AR ATAPS package as additional educational aid to the existing conventional methods?

ANOVA ANALYSIS OF THE RESEARCH QUESTIONS

Research Question 1

This section is comparative in nature and has fifteen questions in total. This section has three parts i.e. Text Book, Current Available Software Packages and AR ATAPS package. The results of the questionnaire are shown in Table 1. Each part has five similar questions all aiming to find the answers for Research Question.

"Can AR ATAPS package help learners in visualizing and understanding the selected difficult engineering problems of the subject matter?"

ANOVA Analysis

Hypothesis: Research Question 1 has been tested against hypothesis that which teaching material (Text Book, Current Package or AR ATAPS package) is preferred by students in terms of understanding the selected difficult engineering problems of the subject matter. Hypothesis made for this question is:

Figure 20. Research question 2

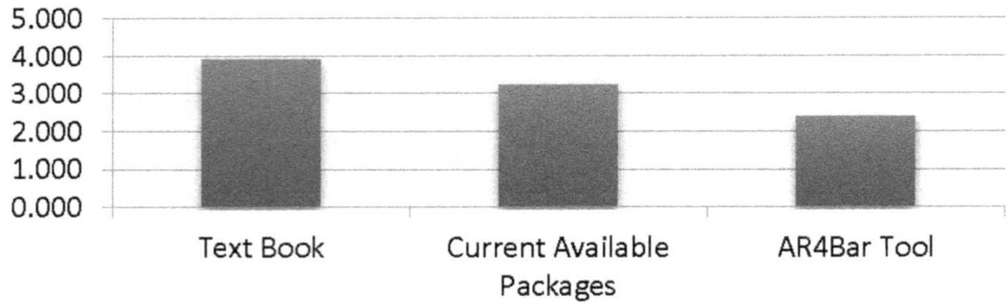

H_0: μ Text Book = μ Current Packages = μ AR ATAPS Package

H_1: μ Text Book ≠ μ Current Packages ≠ μ AR ATAPS Package

Here the number of groups (materials), k=3 and the number of items in each group, n=5. (Note: CF = Correction Factor & SS = Sum of Squares)

CF = ΣX••2/kn = 142.707

Total SS = ΣΣXij2-CF = 5.929

Between materials SS = ΣX•j2/n-CF = 5.238

Within materials = Total SS - Between SS = 0.690

These results and the remaining computations are shown in Table 2 below:

Level of Significance: $\alpha = 0.05$ (5%).

Critical region: F > F0.05 (2, 12) = 3.89

Analysis and Conclusion

Since the calculated value of F = 45.536 is greater than critical value (i.e. F falls in the critical region), the H_0 hypothesis is rejected at 5% level of significance. It can be therefore concluded that the data indicates that "means" of at least two materials are not equal and are significantly different.

Hence it has been concluded that teaching materials i.e. Text Book, Current Software Packages or AR ATAPS package (Figure 19) are preferred differently by students in terms of understanding the selected difficult engineering problems of the subject matter. Efficiency of all the three materials is different according to students.

Referring to research question one, 1 has been assigned to "Strongly Agree" and 5 has been assigned to "Strongly Disagree", so least average would mean (suggests) that more participants agree with the question and highest average would mean (suggests) that more participants disagree with the question.

Table 5. Set of data collected from the survey regarding question 3

Q No.	Text Book (1)	Current Packages (2)	AR ATAPS package (3)	Total
1	**All of the problems and theories can be solved by using (1, 2 and 3)**			
	3.867	3.233	2.333	
2	**I would like to learn solving other problems in Engineering Mechanics topics using (1, 2 and 3)**			
	3.900	3.000	2.033	
3	**The (1, 2 and 3) provides enough information's to solve the exercise questions.**			
	3.200	3.200	2.100	
4	**The way the problem is presented in the (1, 2 and 3) makes it easy to work out the solution.**			
	4.033	3.300	1.867	
5	**(The Figure was attractive and clear), (The screen was attractive and clear) and (The AR ATAPS package user interface was attractive and clear)**			
	3.700	3.167	2.233	
Total	**18.700**	**15.900**	**10.567**	45.167
Mean	**3.740**	**3.180**	**2.113**	

Table 6. Question 3, variance table

Source Of Variation	Sum Of Squares	Degree Of Freedom		Mean Square		F Ratio
Between Tools SS	6.829	k-1 = 3-1	2	s_1^2	3.415	s_1^2/s^2
Within Tools SS	0.601	k(n-1) = 3(5-1)	12	s^2	0.050	68.139
Total SS	7.430	kn-1	14			

Figure 21. Research question 3

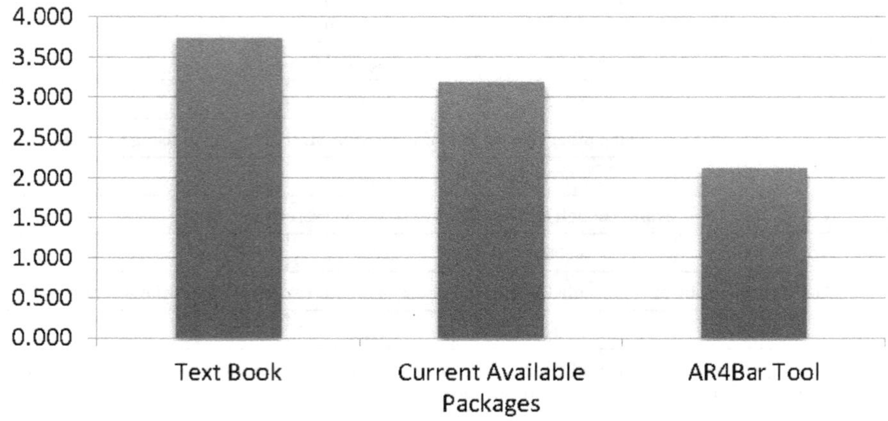

AR ATAPS Package

2.293 < Current Software Packages 3.247<Text Book 3.713

The AR ATAPS package has scored least average indicating that students have agreed more with its efficiency then Current Software Packages and Text Books have been preferred least by scoring most average.

Research Question 2

This section is comparative in nature and has twenty four questions in total. This section has three parts i.e. Text Book, Current Available Software Packages and AR ATAPS package. See for the questionnaire part. The results of the questionnaire are shown in Table 8.3. Each of the three above mentioned materials has eight similar questions all aiming to find the answers for Research Question 2:

"Do virtual problem solving aids with mixed realities and touch to print tangible user interface enhance student learning in engineering?"

ANOVA Analysis

Hypothesis: Research Question 2 has been tested against hypothesis that which teaching material (Text Book, Current Software Packages or AR ATAPS package) is preferred by students in terms of virtual problem solving aids with dynamic illustrations and enhancement of students' learning. Hypothesis made for this question is:

H_0: μ Text Book = μ Current Packages = μ AR ATAPS package

H_1: μ Text Book ≠ μ Current Packages ≠ μ AR ATAPS package

Here the number of groups (tools), k=3 and the number of items in each group, n=8

(Note: CF = Correction Factor & SS = Sum of Squares)

CF = ΣX••2/kn = 240.667

Total SS = ΣΣXij2-CF = 13.091

Between materials SS = ΣX•j2/n-CF = 9.389

Within materials = Total SS - Between SS = 3.702

The remaining computations are shown in the analysis of variance Table 4 below:

Level of Significance: $\alpha = 0.05$ (5%)

Critical region: $F > F_{0.05}(2, 21) = 3.47$

Analysis and Conclusion

Since the calculated value of F = 40.962 is greater than critical value (i.e. F falls in the critical region), the hypothesis H_0 is rejected at 5% level of significance. It has been therefore concluded that the data indicates that "means" of at least two materials are not equal and are significantly different.

Hence it has been concluded that the teaching materials i.e. Text Book, Current Software Packages or AR ATAPS package (Figure 20) all have different level of preferences by students in terms of virtual problem solving aids with dynamic illustrations and enhancement of students' learning. Students have different view for all the three materials.

Referring to research question two, 1 has been assigned to "Strongly Agree" and 5 has been assigned to "Strongly Disagree" so least average would mean (suggests) that more participants agree with the question and highest average would mean (suggests) that more participants disagree with the question.

AR ATAPS Package

2.375 < Current Software Packages 3.221 < Text Book3.904

The AR ATAPS package has scored least average indicating that students have agreed more with its efficiency than Current Software Packages and Text Books have been preferred least by scoring most average.

Research Question 3

This section is comparative in nature and has fifteen questions in total. This section has three parts i.e. Text Book, Current Available Software Packages and AR ATAPS package. The results of the questionnaire are shown in Table 5. Each part has five similar questions all aiming to find the answers for Research Question 3:

"Do the learners accept the AR ATAPS package as additional educational aid to the existing conventional methods?"

ANOVA Analysis

Hypothesis: Research Question 3 has been tested against hypothesis that which teaching material (Text Book, Current Software Package or AR ATAPS Package) is preferred by students in terms additional educational aid. Hypothesis made for this question is:

H_0: μ Text Book = μ Current Packages = μ AR ATAPS package

H_1: μ Text Book ≠ μ Current Packages ≠ μ AR ATAPS package

Here the number of groups (materials), k=3 and the number of items in each group, n=5

(Note: CF = Correction Factor & SS = Sum of Squares)

CF = ΣX••2/kn = 136.002

Total SS = ΣΣXij2-CF = 7.430

Between materials SS = ΣX•j2/n-CF = 6.829

Within material = Total SS - Between SS = 0.601

These results and the remaining computations are shown in the analysis of variance Table 6 below:

Level of Significance: $\alpha = 0.05$ (5%)

Critical region: $F > F_{0.05}$ (2, 21) = 3.89

Analysis and Conclusion

Since the calculated value of F = 68.139 is greater than critical value (i.e. F falls in the critical region), the hypothesis H_0 is rejected at 5% level of significance. It can therefore be concluded that the data indicates that "means" of at least two materials are not equal and are significantly different.

Hence it has been concluded that teaching materials (Text Book, Current Software Packages or AR ATAPS Package) are differently preferred by students in terms of additional educational aid. Students have different opinion for all the three materials.

Referring to research question three, 1 has been assigned to "Strongly Agree" and 5 has been assigned to "Strongly Disagree" so least average would mean (suggest) that more participants agree with the question and highest average would mean (suggest) that more participants disagree with the question (Figure 21).

AR ATAPS Package

2.113 < Current Software Packages 3.180 < Text Book 3.740

The AR ATAPS package has scored least average indicating that students have agreed more with its efficiency then Current Software Packages and Text Books have been preferred least by scoring most average.

SUMMARY

In summary, this Chapter has discussed about AR environments. The characteristics, architecture, interaction techniques, general AR and engineering applications were also discussed in details followed by benefits.

As a whole, AR systems and applications have great potential and have been widely used across a wide range of courses to promote learning. In addition AR is the perfect technology for groups of learners to thrive in a project and problem solving based learning environment. Each learner can use their AR capable device to build or help solve problems. However, due to the different nature of each field of study, the degree to which computers can be used in teaching varies greatly.

In terms of using a particular AR Toolkit and designing user interfaces, this may depend on the purpose and nature of the application under development.

If the developer fails to design an appropriate user interface, this may not promote learning and hamper the application. Consequently users may reject its usage as a learning software tool.

While there are various approaches of user interface design, few of them concentrate basically on the users and the learning processes. These approaches might be practical but very complex to be useful for practitioners of interface design or not strongly rooted in the theory of learning. In addition, there is no approach founded in the literature which is specified for instructional strategies, learning packages or for the methods that are relevant to this field. More standards and researched guidelines may help in designing better user interfaces by applying suitable design principles and methods, i.e. interfaces that are easier to learn and use.

A brief theory of 4 link bar mechanisms was provided. These mechanisms are important for students to understand in order to analyze and visualize problems pertaining to 4 link bar mechanisms. If the students fail to understand these mechanisms, they will not be able to comprehend and apply the correct equations and formulas to solve the problems. As such the ATAPS software has been designed and structured in such a way that it should help the students perform the computations and analyze any of the 4 link-bar problems based on the Grashof's law and the values input by the user.

The ATAPS software also models the 4 bar linkages in 3D real time environment and forms on the screen and shows the correct orientations and animation of the linkages. Consequently students are able to view the 3D models from any angle to examine the 3D model while the model and analysis is being simultaneously updated till the user terminates it. Such animated movements cannot be shown in the textbook due to its static nature. Further description of the AR ATAPS software provided insights of how it had helped students to visualize and understand the problem better.

The AR ATAPS software was developed to assist engineering students in studying mechanics courses, and it was designed as a Computer Aided Tool for 3D modeling and analysis of the selected engineering problems. This problem was previously found to be difficult for students to understand during normal classroom lectures. Therefore the AR ATAPS was developed to familiarize students with the concepts of solving Engineering Mechanics Dynamics problem.

The evaluation of this software tool was performed using a comparative test on Text Books, Current Available Packages and AR ATAPS on a sample that contains 30 UNITEN engineering students, ANOVA analysis was used to analyze the results of the survey. The survey has six sections and the evaluation has also been part of the six sections. The first three sections have been analyzed by simple percentage analysis and the last three sections have been analyzed by ANOVA.

The first section is related to Experience with Computing. The data from this section shows that most of the respondents have access to computer; they are familiar with usage of computer for educational purpose and are currently using some Computer Software Packages for engineering subjects.

Second section is about Learning Experience of students at UNITEN during the Mechanical Courses. Statistics shows that most of the participants find the course and teaching material unsatisfactory whereas few are satisfied with the way of teaching.

The third section is concerned with Computing Knowledge and AR ATAPS package. The results for this section could be concluded as most of the participants agreed with the efficiency of the AR ATAPS package and it's been found helpful, beneficial and no negative points related to the AR ATAPS package have come across during the analysis of the data.

The fourth section which is related to Research (Question 1), and is analyzed with the help of ANOVA has been concluded that AR ATAPS package is preferred most by students in terms of understanding the selected difficult engineering problems of the subject matter. Refer to section 8.16.1.

It has been concluded in section 8.16.2 that is related to the research (Question 2), that AR ATAPS package got the highest preference by students in terms of virtual problem solving aids with dynamic illustrations and enhancement of students' learning.

In section 8.16.3 that is related to research (Question 3), AR ATAPS package has scored least average indicating that students have agreed more with its efficiency then Current Software Packages and Text Books have been preferred least by scoring most average. Which indicate that AR ATAPS package is most preferred and accepted by the students as additional learning aid.

ACKNOWLEDGMENT

The author would like to acknowledge the significant contributions of Waleed Maqableh a PhD in ICT student who implemented the AR software and performed analysis of the data for the research.

REFERENCES

Alcañiz, M., Contero, M., & Pérez-López, D. C. (2010). *Augmented Reality Technology for Education. InstitutoInteruniversitario de Investigación en BioingenieríayTecnologíaOrientada al SerHumano (I3BH).* Universidad Politécnica de Valencia Spain.

Ariyana, Y., & Wuryandari, A. I. (2012). Basic 3D Interaction Techniques in Augmented Reality. *International Conference on System Engineering and Technology*. 10.1109/ICSEngT.2012.6339281

Azuma, Balliot, Behringer, Feiner, Julier, & MacIntyre. (2001). *Recent Advances in Augmented Reality Computers & Graphics*. Academic Press.

Azuma, R. (1997). A Survey of Augmented Reality. *Presence (Cambridge, Mass.)*, *6*(4), 355–385. doi:10.1162/pres.1997.6.4.355

Azuma, R. T., Baillot, Y., Behringer, R., Feiner, S. K., Julier, S., & MacIntyre, B. (2001, November-December). Recent advances in augmented reality. *IEEE Computer Graphics and Applications*, *21*(6), 34–47. doi:10.1109/38.963459

Bandyopadhyay, D., Raskar, R., & Fuchs, H. (2001). Dynamic shader lamps: Painting on movable objects. *Proc. ISMAR '01 (IEEE and ACM International Symposium on Augmented Reality),* 207–216.

Bell, J. T., & Fogler, H. S. (1995). The Investigation and Application of Virtual Reality as an Educational Tool. *Proceedings of American Society for Engineering Education Annual Conference 1995.*

Billinghurst, M. (2008). *Usability testing of augmented/mixed reality systems*. International Conference on Computer Graphics and Interactive Techniques (ACM SIGGRAPH ASIA 2008 courses), Singapore.

Billinghurst, M. (2002). Augmented Reality in Education. *New Horizons for Learning*. Retrieved from http://www. newhorizons. org/strategies/technology/ billinghurst. htm

Billinghurst, M., Kato, H., & Poupyrev, I. (2001). The MagicBook: A Transitional AR Interface. *Computers & Graphics*, *25*, 745–753. doi:10.1016/ S0097-8493(01)00117-0

Billinghurst, M., Kato, H., & Poupyrev, I. (2001). *Collaboration with tangible augmented reality interfaces*. Paper presented at HCI International 2001.

Bogen, M., Wind, J., & Giuliano, A. (2006). Augmented Reality in School Environments. EC-TEL 2006. *LNCS*, *4227*, 709–714.

Bostanci, E., & Kanwal, N. (2010). Tracking Methods for Augmented Reality. *The 3rd International Conference on Machine Vision (ICMV 2010).*

Butchart, B. (2011). Augmented Reality for Smartphones A Guide for developers and content publishers. *JISC Observatory, Observing trends in Innovation.*

Byrne, C. (1996). *Water on Tap: The Use of Virtual Reality as an Educational Tool* (PhD thesis). University of Washington, College of Engineering.

Chen, Y.-C. (2006). A Study of Comparing the Use of Augmented Reality and Physical Models in Chemistry Education. VRCIA 2006, Hong Kong. doi:10.1145/1128923.1128990

Chen, Feng, Liu, Siyuan, Zhongning, & Huang. (2011). *Application of Augmented Reality in Engineering Graphics Education*. IEEE.

Coe, U. (2013). *Vision & Mission.* Retrieved March 15, 2013, from Universiti Tenaga Nasional, UNITEN: http://www.uniten.edu.my/newhome/content_list.asp?ContentID=3101

Cohen, P. R. (1997, December). A touring machine: Prototyping 3D mobile augmented reality systems for exploring the urban environment. *Personal and Ubiquitous Computing*, *1*(4), 74–81.

Cooper, N., Keatley, A., Dahlquist, M., Mann, S., Slay, H., Zucco, J., . . . Thomas, B. H. (2004). *Augmented Reality Chinese Checkers. In International Conference on Advances in Computer Entertainment Technology*. Academic Press.

deLandgraaf, W. A. (2003/2004). *Interaction between users and Augmented Reality systems: Human-Computer Interaction of the future*. Vrije Universiteit Amsterdam.

Dolz, J. (2012, May 17). *Markerless Augmented Reality.* Retrieved March 15, 2013, from AR Lab (Augumented Reality): http://www.arlab.com/blog/markerless-augmented-reality/

Dünser, A., & Hornecker, E. (2007). Lessons from an AR Book study. TEI'07, Baton Rouge, LA. doi:10.1145/1226969.1227006

Durlach, N., Allen, G., Darken, R., Garnett, R. L., Loomis, J., Templeman, J., & von Wiegand, T. E. (2000). Virtual environments and the enhancement of spatial behavior: Towards a comprehensive research agenda. *Presence (Cambridge, Mass.)*, *9*(6), 593–615. doi:10.1162/105474600300040402

EduCause. (2005, September). *7 things you should know about Augmented Reality*. Retrieved April 12, 2013, from EduCause Learning Initiative; Advancing Learning through IT innovation: www.educause.edu/eli

Eike, F. A. & Fotis, L. (2010). *Using Augmented Reality as a Medium to Assist Teaching in Higher Education*. Euro Graphics Association.

Ferrari, V. (2001). Markerless augmented reality with a real-time affine region tracker. *Augmented Reality, 2001. Proceedings. IEEE and ACM International Symposium on Date of Conference*, 87-96. 10.1109/ISAR.2001.970518

Fitzgibbon & Zisserman. (2000). Markerless tracking using planar structures in the scene. *Procs. of the Instep. on Augmented Reality*.

Fitzmaurice, G. W. (1993). Situated information spaces and spatially aware palmtop computers. *Communications of the ACM*, *36*(7), 39–49. doi:10.1145/159544.159566

Griffin, D. (2013). *Augmented Reality Learning Environments – A posthuman pedagogy*. Retrieved May 12, 2013, from www.danielgriffin.net: http://www.danielgriffin.net/cms/node/81

Guralnick, D. A. (2006). User Interface Design for Effective, Engaging E-Learning. In *Proceedings of the International Conference on e-Learning* (p. 171). ACL.

Henderson, S., & Feiner, S. (2011). Exploring the Benefits of Augmented Reality Documentation for Maintenance and Repair. *IEEE Transactions on Visualization and Computer Graphics*, *17*(10).

Henderson, S., & Steven, F. (2007). *Augmented Reality for Maintenance and Repair (ARMAR)*. Technical Report AFRL-RH-WP-TR-2007-0112, United States Air Force Research Lab.

Höllerer, T. H., & Feiner, S. K. (2004). Mobile Augmented Reality. In H. A. Karimi & A. Hammad (Eds.), *Telegeoinformatics: Location-Based Computing and Services* (pp. 392–421). CRC Press.

Huber, G., & Dietmaier, P. (2010). Four Bar: An educational synthesis and designtool for four bar mechanisms. *Proceedings of the 10th World Congresson the Theory Machines and Mechanisms*, 20–24.

Hughes, C. E., Stapleton, C. B., Hughes, D. E., & Smith, E. M. (2005, December). Mixed reality in education, entertainment, and training. *IEEE Computer Graphics and Applications*, *25*(Nov), 24–30. doi:10.1109/MCG.2005.139 PMID:16315474

Ishii, H., & Ullmer, B. (1997). Tangible bits: towards seamless interfaces between people, bits and atoms. In *CHI '97: Proceedings of the SIGCHI conference on Human factors in computing systems*. New York. ACM Press. 10.1145/258549.258715

Jang, S. H., & Hudson-Smith, A. (2012). Exploring Mobile Augmented Reality Navigation System for Pedestrians. *Proceedings of the GIS Research UK 20th Annual Conference GISRUK 2012*.

Kaufmann, H. (2003). *Collaborative Augmented Reality in Education*. IMAGINA 2003 Conference CD, Monte Carlo, Monaco.

Kaufmann, H., & Schmalstieg, D. (2003). Mathematics and geometry education with collaborative augmented reality. Computers & Graphics, 27(3), 339-345. doi:10.1016/S0097-8493(03)00028-1

Kerawalla, L., Luckin, R., Seljefot, S., & Woolard, A. (2006). Making it real: Exploring the potential of augmented reality for teaching primary school science. *Virtual Reality (Waltham Cross)*, *10*(3-4), 163–174. doi:10.100710055-006-0036-4

Kinshuk, R. (1996). *Computer Aided Learning for Entry accountancy students* (PhD Thesis). De Montfort University.

Kinshuk, R., & Patel, A. (2003). Optimizing Domain Knowledge Representation with Multimedia Objects. Learning and Teaching with Technology - Principles and practices, London, UK.

Klopfer, E., Squire, K., & Jenkins, H. (2002). *Environmental Detectives PDAs as a Window intoa Virtual Simulated World*. Paper presented at International Workshop on Wireless and Mobile Technologies in Education, Vaxjo, Sweden. 10.1109/WMTE.2002.1039227

Kurt, S., & Eric, K. (2004). Gmented Reality Simulations on Handheld Computers. School of Education, UW Madison.

Larsen, Buchholz, Brosda, & Bogner. (2011). Evaluation of a portable and interactive Augmented reality learning system by teachers and students. *Proceedings of the "Science Center To Go" Workshops.*

Lester. (2012, October 31). *Readers' Choice Awards 2012 – Results.* Retrieved March 10, 2013, from Augmented Planet - The Augmented Reality Blog: http://www.augmentedplanet.com/2012/10/readers-choice-awards-2012-results/

Liarokapis, F., Mourkoussis, N., Petridis, P., Rumsey, S., Lister, P. F., & White, M. (2002). *An Interactive Augmented Reality System for Engineering Education. 3rd Global Congress on Engineering Education*, Glasgow, UK.

Liarokapis, F., Mourkoussis, N., White, M., Darcy, J., Sifniotis, M., Petridis, P., ... Lister, P. F. (2007). Augmented Reality and Web 3D to Support Engineering Education. *Proceedings of the Engineering and Technology Education Conference*, 24-28.

Liarokapis, F., Petridis, P., Lister, P. F., & White, M. (2002). Multimedia Augmented Reality Interface for E-Learning (MARIE). *World Transactions on Engineering and Technology Education*, *1*(2), 173–176.

Loftin, B., Engelberg, M., & Benedetti, R. (1993). Applying virtual reality in education: A prototypical virtual physics laboratory. *Proceedings of IEEE 1993 Symposium on Research Frontiers in Virtual Reality*, 67-74. 10.1109/VRAIS.1993.378261

Ludwig, C., & Reimann, C. (2005). Augmented reality: Information at focus. *Cooperative Computing & Communication Laboratory, 4*(1).

Maier, P., Tönnis, M., & Klinker, G. (2009). Augmented Reality for teaching spatial relations. Conference of the International Journal of Arts & Sciences.

Makina & Salam. (2011). User Interface and Interaction Design Considerations for Collaborative Learning Using Augmented Reality Learning Object. *ICSECS 2011, Part I, CCIS 179*, 179–187.

Mantovani, F. (2001). VR Learning: Potential and Challenges for the Use of 3D Environments in Education and Training. Towards Cyber Psychology: Mind. *Cognitions and Society in the Internet Age*, *2*, 207–225.

Maqableh, W. F. (2012, November). Preliminary Evaluation of using Augmented Reality as a Dynamic Simulation Learning Tool for Linkage Mechanisms. *International Journal of Computers and Applications*, *58*(4), 19–23. doi:10.5120/9270-3460

Maqableh, W. F., & Sidhu, M. S. (2010) From Boards to Augmented Reality Learning. *International Conference on Information Retrieval and Knowledge Management*, 184-187.

Marchese, F. T. (2008, September 6). *AR System Characteristics in "The User Interface from Front to Back; Augmented Reality"*. Retrieved March 12, 2013, from: http://csis.pace.edu/~marchese/DPS/Lect3/dpsl3.html

Martín-Gutiérrez, Saorín, Contero, & Alcañiz. (2010). AR_Dehaes: An Educational Toolkit Based on Augmented Reality Technology for Learning Engineering Graphics. *10th IEEE International Conference on Advanced Learning Technologies*. DOI 10.1109/ICALT.2010.45

Martín-Gutiérrez, J., & Contero, M. (2011). Mixed Reality for Learning Standard Mechanical Elements. *11th IEEE International Conference on Advanced Learning Technologies*. 10.1109/ICALT.2011.117

Martín-Gutiérrez, J., Saorín, J. L., Contero, M., Alcañiz, M., Pérez López, D. C., & Ortega, M. (2010). Design and Validation of an Augmented Reality for Spatial Abilities Development in Engineering Students. *Computer Graphics, 34*(1), 77–91. doi:10.1016/j.cag.2009.11.003

Matcha, W., & Awang Rambli, D. R. (2012). Design consideration for Augmented Reality book-based application for collaborative learning environment. *Computer & Information Science (ICCIS), International Conference on, 2*, 1123-1126. 10.1109/ICCISci.2012.6297194

O'Brien, H. L., & Toms, E. G. (2005). Engagement as Process in Computer-Mediated Environments. *Proceedings of ASIS&T*.

O'Malley, C. (1990). *Interfaces issues for guided discovery learning environments. Guided Discovery Tutoring*. London: Paul Chapman.

Oh, S., & Woo, W. (2008). *ARGarden: Augmented Edutainment System with a Learning Companion*. Springer-Verlag Berlin Heidelberg.

Petersen, T. (2008). *A Comparison of 2D-3D Pose Estimation Methods*. Aalborg University.

Pinhanez, C. S. (2001). The everywhere displays projector: A device to create ubiquitous graphical interfaces. *UBICOMP '01 (International Conference on Ubiquitous Computing)*, 315–331. 10.1007/3-540-45427-6_27

Piumsomboon, T., Clark, A., & Billinghurst, M. (2011). Physically-based Interaction for Tabletop Augmented Reality Using a Depth-sensing Camera for Environment Mapping. Proc. Image and Vision Computing New Zealand (IVCNZ-2011), 161-166.

Potter, G. (2011). *Augmented Reality and Mobile Technologies. Models for Interdisciplinary Mobile Learning*. IGI Global. doi:10.4018/978-1-60960-511-7.ch012

Ramirez, M. (2011). *The Benefits of using Augmented Reality in Education*. Retrieved March 14, 2013, from SCARLET (Special Collections using Augmented Reality to Enhance Learning and Teaching): http://teamscarlet.wordpress.com/2011/08/09/the-benefits-of-using-augmented-reality-in-education/

Robert, D. (1994). *The Development, Formative and Summative Evaluation of a Computer Multimedia Tutorial: A Case Study* (PhD thesis). Walden University.

Rolland, J., Holloway, R., & Fuchs, H. (1994). A Comparison of Optical and Video See-Through Head-Mounted Displays. *SPIE Proceedings volume 2351: Telemanipulator and Telepresence Technologies*, 293-307.

Salisbury, M. W., Hendrickson, J. H., Lammers, T. L., Fu, C., & Moody, S. A. (1990). Talk and draw: Bundling speech and graphics. *IEEE Computer.*, *23*(8), 59–65. doi:10.1109/2.56872

Schmalstieg, D., & Wagner, D. (2007). *Mobile Phones as a Platform for Augmented Reality*. Graz: University of Technology.

Seichter, H., Grasset, R., Looser, J., & Billinghurst, M. (2009). Multitouch Interaction for Tangible User Interfaces. In *ISMAR 2009*. Orlando, FL: ACM Press. doi:10.1109/ISMAR.2009.5336455

Selker, T. (2008). Touching the future. *Communications of the ACM*, *51*(12), 14–16. doi:10.1145/1409360.1409366

Shelton, B. E., & Hedley, N. R. (2002). Using augmented reality for teaching earth-sun relationship to undergraduate geography students. In *The First IEEE International Augmented Reality Toolkit Workshop* (pp. 1-8). Darmstadt, Germany: IEEE. 10.1109/ART.2002.1106948

Sidhu. (2007). *Technology-Assisted Problem Solving for Engineering Applications* (PhD thesis). University Malaya.

Squire & Jan. (2008). *Wherever You Go, There You Are: Place-Based Augmented Reality Games For Learning*. Academic Press.

Sterling, B. (2010). *Research in Educational AR*. Retrieved March 02, 2013, from http://educationar.wikispaces.com/Research+in+Educational+AR

Taxén, G., & Naeve, A. (2001). CyberMath: Exploring Open Issues in VR-Based Learning. In *Proceedings of SIGGRAPH 2001*, (pp. 49-51). Los Angeles, CA: Association for Computing Machinery.

TeachThought. (2012). *Augmented Reality In Education? Here Are 20 Examples*. Retrieved March 14, 2013, from TeachThought 2012: http://www.teachthought.com/technology/augmented-reality-in-education-here-are-20-examples/

UNITEN. (2013). *Introduction*. Retrieved March 12, 2013, from Universiti Tenaga Nasional, UNITEN: http://www.uniten.edu.my/newhome/content_list.asp?contentid=1877

Valentini, P. P. (2010). Interactive Multibody Simulation in Augmented Reality. Journal of Theoretical and Applied Simulation, 48(3), 733-750.

Vera, Gimeno, Coma, & Fernández. (2011). Augmented Mirror: Interactive Augmented Reality System Based on Kinect. *LNCS, 6949*, 483–486.

Wagner, D., Reitmayr, G., Mulloni, A., Drummond, T., & Schmalstieg, D. (2008). Pose Tracking from Natural Features on Mobile Phones. ISMAR'08. doi:10.1109/ISMAR.2008.4637338

Wilson, A. D. (2004). Touchlight, An imaging touch screen and display for gesture-based interaction. *Proc. ICMI '04 (International conference on Multimodal interfaces)*, 69–76.

Winn, W. (1997). *The Impact of Three-Dimensional Immersive Virtual Environments on Modern Pedagogy*. University of Washington, Human Interface Technology Laboratory, Technical Report R-97-15.

Wolfram, S. (2002). A New kind of science. Champaign, IL: Wolfram Media.

Xu, D. (2005). Tangible User Interface for Children an Overview. *Proc. of the UCLAN Department of Computing Conference*.

Zhou, F., Duh, H. L., & Billinghurst, M. (2008). Trends in augmented reality traching, interaction and display: A review of ten years in ISMAR. In *Mixed and Augmented Reality, ISMAR 7th IEE/ACM International Symposium* (pp. 193-202). Cambridge, UK: IEEE.

KEY TERMS AND DEFINITIONS

ANOVA: One-way or two-way refers to the number of independent variables (IVs) in your Analysis of Variance test. One-way has one independent variable (with 2 levels) and two-way has two independent variables (can have multiple levels).

Grashof's Law: The law states that for a four-bar linkage system, the sum of the shortest and longest link of a planar quadrilateral linkage is less than or equal to the sum of the remaining two links, then the shortest link can rotate fully with respect to a neighbouring link.

Marker-Less Augmented Reality: Relies in natural features of the image or object to be tracked, like the edges, corners, or textures.

Touch Interactions: This kind of interaction can directly be applied to graphical objects of interest without the need for special-purpose devices that users may need to operate in other forms of human-computer interaction.

Related Readings

To continue IGI Global's long-standing tradition of advancing innovation through emerging research, please find below a compiled list of recommended IGI Global book chapters and journal articles in the areas of online learning, blended learning, and technological pedagogical content knowledge. These related readings will provide additional information and guidance to further enrich your knowledge and assist you with your own research.

Ackerman, G. L. (2017). Strategies to Increase Technology Acceptance. In M. Grassetti & S. Brookby (Eds.), *Advancing Next-Generation Teacher Education through Digital Tools and Applications* (pp. 1–18). Hershey, PA: IGI Global. doi:10.4018/978-1-5225-0965-3.ch001

Adeyele, V., & Aladejana, F. (2019). Comparative Effectiveness of Interactive Multimedia, Simulation Games, and Blended Learning on Science Performance of Learners With Special Needs. In J. Keengwe (Ed.), *Handbook of Research on Blended Learning Pedagogies and Professional Development in Higher Education* (pp. 340–356). Hershey, PA: IGI Global. doi:10.4018/978-1-5225-5557-5.ch019

Ahuja, S. (2017). Using the Flipped Classroom to Improve Knowledge Creation of Master's-Level Students in Engineering. In T. Kidd & L. Morris Jr., (Eds.), *Handbook of Research on Instructional Systems and Educational Technology* (pp. 326–339). Hershey, PA: IGI Global. doi:10.4018/978-1-5225-2399-4.ch028

Asunka, S. (2017). "We Had a Blast!": An Empirical Affirmation of Blended Learning as the Preferred Learning Mode for Adult Learners. *International Journal of Mobile and Blended Learning*, *9*(3), 37–53. doi:10.4018/IJMBL.2017070104

Bailey, E. K. (2017). Applying Learning Theories to Computer Technology Supported Instruction. In M. Grassetti & S. Brookby (Eds.), *Advancing Next-Generation Teacher Education through Digital Tools and Applications* (pp. 61–81). Hershey, PA: IGI Global. doi:10.4018/978-1-5225-0965-3.ch004

Bernal, E. V., & Farmer, L. S. (2018). Evaluating a Professional Development Program for Course Redesign With Technology: The Faculty End-User Experience. In R. Roscoe, S. Craig, & I. Douglas (Eds.), *End-User Considerations in Educational Technology Design* (pp. 316–336). Hershey, PA: IGI Global. doi:10.4018/978-1-5225-2639-1.ch015

Bosch, C., Mentz, E., & Reitsma, G. M. (2019). Integrating Cooperative Learning into the Combined Blended Learning Design Model: Implications for Students' Intrinsic Motivation. *International Journal of Mobile and Blended Learning*, *11*(1), 58–73. doi:10.4018/IJMBL.2019010105

Brown, L. M. (2019). Today's Demands on Education. In D. Williams & N. Harkness (Eds.), *Diverse Learning Opportunities Through Technology-Based Curriculum Design* (pp. 42–68). Hershey, PA: IGI Global. doi:10.4018/978-1-5225-5519-3.ch003

Brown, L. M. (2019). What Is Technology? In D. Williams & N. Harkness (Eds.), *Diverse Learning Opportunities Through Technology-Based Curriculum Design* (pp. 16–40). Hershey, PA: IGI Global. doi:10.4018/978-1-5225-5519-3.ch002

Casanova, D., & Moreira, A. (2017). A Model for Discussing the Quality of Technology-Enhanced Learning in Blended Learning Programmes. *International Journal of Mobile and Blended Learning*, *9*(4), 1–20. doi:10.4018/IJMBL.2017100101

Cheng, H. (2017). A TL-TPACK Model on CSL Pre-Service Teachers' Competencies of Online Instruction. In C. Lin, D. Zhang, & B. Zheng (Eds.), *Preparing Foreign Language Teachers for Next-Generation Education* (pp. 198–225). Hershey, PA: IGI Global. doi:10.4018/978-1-5225-0483-2.ch011

Collins, P. K. (2017). Strategies to Support the Faculty Adoption of Technology for Student Success Initiatives. In T. Kidd & L. Morris Jr., (Eds.), *Handbook of Research on Instructional Systems and Educational Technology* (pp. 392–399). Hershey, PA: IGI Global. doi:10.4018/978-1-5225-2399-4.ch033

Conger, S., Krauss, K. E., & Simuja, C. (2017). New Pedagogical Approaches with Technologies. *International Journal of Technology and Human Interaction*, *13*(4), 62–76. doi:10.4018/IJTHI.2017100105

Connally, P. N., & Morris, L. R. (2017). The Impact of Emerging Technology on Leadership Development. In T. Kidd & L. Morris Jr., (Eds.), *Handbook of Research on Instructional Systems and Educational Technology* (pp. 400–408). Hershey, PA: IGI Global. doi:10.4018/978-1-5225-2399-4.ch034

Cressey, J. (2017). Universal Design for Learning: Using UDL to Make Teacher Education More Accessible and Inclusive for All. In M. Grassetti & S. Brookby (Eds.), *Advancing Next-Generation Teacher Education through Digital Tools and Applications* (pp. 38–59). Hershey, PA: IGI Global. doi:10.4018/978-1-5225-0965-3.ch003

Demir, K. (2017). Mobile Learning. In T. Kidd & L. Morris Jr., (Eds.), *Handbook of Research on Instructional Systems and Educational Technology* (pp. 291–298). Hershey, PA: IGI Global. doi:10.4018/978-1-5225-2399-4.ch025

Dillard, M. S. (2019). Views of Technology Importance for Developmental Students at the Community College Level. In D. Williams & N. Harkness (Eds.), *Diverse Learning Opportunities Through Technology-Based Curriculum Design* (pp. 196–216). Hershey, PA: IGI Global. doi:10.4018/978-1-5225-5519-3.ch009

Dixon, J., & Shen, L. (2019). Technology-Integrated Curriculum and Students' Academic Performance. In D. Williams & N. Harkness (Eds.), *Diverse Learning Opportunities Through Technology-Based Curriculum Design* (pp. 96–131). Hershey, PA: IGI Global. doi:10.4018/978-1-5225-5519-3.ch005

Dreon, O., Shettel, J., & Bower, K. M. (2017). Preparing Next Generation Elementary Teachers for the Tools of Tomorrow. In M. Grassetti & S. Brookby (Eds.), *Advancing Next-Generation Teacher Education through Digital Tools and Applications* (pp. 143–159). Hershey, PA: IGI Global. doi:10.4018/978-1-5225-0965-3.ch008

Evans, N. D. (2019). Connecting Higher Education Learning Spaces in a Blended Zululand Teaching and Learning Ecology. In J. Keengwe (Ed.), *Handbook of Research on Blended Learning Pedagogies and Professional Development in Higher Education* (pp. 177–195). Hershey, PA: IGI Global. doi:10.4018/978-1-5225-5557-5.ch010

Eyyam, R. (2017). The Effect of Technology-Enhanced Classrooms in Middle School Education. In M. Grassetti & S. Brookby (Eds.), *Advancing Next-Generation Teacher Education through Digital Tools and Applications* (pp. 119–142). Hershey, PA: IGI Global. doi:10.4018/978-1-5225-0965-3.ch007

Farmer, L. S. (2019). The Role of Librarians in Blended Courses. In J. Keengwe (Ed.), *Handbook of Research on Blended Learning Pedagogies and Professional Development in Higher Education* (pp. 122–138). Hershey, PA: IGI Global. doi:10.4018/978-1-5225-5557-5.ch007

Ficarra, L. R., & Chapin, D. A. (2019). The Role of Course Management Systems (CMS) in Addressing Universal Design for Learning (UDL) in College Classrooms. In J. Keengwe (Ed.), *Handbook of Research on Blended Learning Pedagogies and Professional Development in Higher Education* (pp. 60–78). Hershey, PA: IGI Global. doi:10.4018/978-1-5225-5557-5.ch004

Floyd, K. K., & Shambaugh, N. (2017). Instructional Design for Simulations in Special Education Virtual Learning Spaces. In T. Kidd & L. Morris Jr., (Eds.), *Handbook of Research on Instructional Systems and Educational Technology* (pp. 202–215). Hershey, PA: IGI Global. doi:10.4018/978-1-5225-2399-4.ch018

Fraga, L. M., Foster, S. L., & Falcon, L. (2019). Model for My Students?: I'm Not Sure How to Integrate Technology Either! In J. Keengwe (Ed.), *Handbook of Research on Blended Learning Pedagogies and Professional Development in Higher Education* (pp. 274–289). Hershey, PA: IGI Global. doi:10.4018/978-1-5225-5557-5.ch015

Friesem, Y., Jennings, B. J., & Prest, C. (2017). Let It Go: A Journey toward Elementary Student-Driven Media Production Aligned with the CCSS. In M. Grassetti & S. Brookby (Eds.), *Advancing Next-Generation Teacher Education through Digital Tools and Applications* (pp. 245–261). Hershey, PA: IGI Global. doi:10.4018/978-1-5225-0965-3.ch013

Ghazal, S., Aldowah, H., Umar, I., & Bervell, B. (2018). Acceptance and Satisfaction of Learning Management System Enabled Blended Learning Based on a Modified DeLone-McLean Information System Success Model. *International Journal of Information Technology Project Management*, *9*(3), 52–71. doi:10.4018/IJITPM.2018070104

Ghazanfarian, J., Khavasi, E., Yousefi, H., Amiraslanpour, M., Teymouri, S., & Bayat, R. (2018). Modern Online Learning Tools Over the Platform of Virtual/Augmented Reality. In A. Kumar (Ed.), *Optimizing Student Engagement in Online Learning Environments* (pp. 101–126). Hershey, PA: IGI Global. doi:10.4018/978-1-5225-3634-5.ch005

Giovannini, J. M. (2017). Technology Integration in Preservice Teacher Education Programs: Research-based Recommendations. In M. Grassetti & S. Brookby (Eds.), *Advancing Next-Generation Teacher Education through Digital Tools and Applications* (pp. 82–102). Hershey, PA: IGI Global. doi:10.4018/978-1-5225-0965-3.ch005

Goyal, M., & Krishnamurthy, R. (2018). Optimizing Student Engagement in Online Learning Environments: Intuitionistic Fuzzy Logic in Student Modeling. In A. Kumar (Ed.), *Optimizing Student Engagement in Online Learning Environments* (pp. 187–219). Hershey, PA: IGI Global. doi:10.4018/978-1-5225-3634-5.ch009

Graesser, A. C., Nixon, T. M., Hampton, A. J., Franklin, S. E., & Love, J. B. (2018). Via: Using GOMS to Improve Authorware for a Virtual Internship Environment. In R. Roscoe, S. Craig, & I. Douglas (Eds.), *End-User Considerations in Educational Technology Design* (pp. 205–216). Hershey, PA: IGI Global. doi:10.4018/978-1-5225-2639-1.ch010

Gretter, S., & Gondra, A. (2017). Teaching Foreign Languages in the Twenty-First Century: Lessons from Spanish Hybrid Education. In C. Lin, D. Zhang, & B. Zheng (Eds.), *Preparing Foreign Language Teachers for Next-Generation Education* (pp. 92–113). Hershey, PA: IGI Global. doi:10.4018/978-1-5225-0483-2.ch006

Grosbois, M. M., & Sarré, C. G. (2017). Learning to Teach for Next-Generation Education: A Careful Blend of Action and Reflection. In C. Lin, D. Zhang, & B. Zheng (Eds.), *Preparing Foreign Language Teachers for Next-Generation Education* (pp. 153–174). Hershey, PA: IGI Global. doi:10.4018/978-1-5225-0483-2.ch009

Güzeldereli, E. A., Kose, U., & Tüfekci, A. (2017). Human Performance Technology and the Effects on Web-Based Instruction Performance Efficiency. In T. Kidd & L. Morris Jr., (Eds.), *Handbook of Research on Instructional Systems and Educational Technology* (pp. 89–103). Hershey, PA: IGI Global. doi:10.4018/978-1-5225-2399-4.ch009

Gyaase, P. O., Gyamfi, S. A., & Kuranchie, A. (2019). Gauging the E-Readiness for the Integration of Information and Communication Technology Into Pre-Tertiary Education in Ghana: An Assessment of Teachers' Technological Pedagogical Content Knowledge (TPACK). *International Journal of Information and Communication Technology Education, 15*(2), 1–17. doi:10.4018/IJICTE.2019040101

Halvorsen, A. D. (2017). Divergent Teacher Viewpoints of Technology Integration in the Language Classroom. In C. Lin, D. Zhang, & B. Zheng (Eds.), *Preparing Foreign Language Teachers for Next-Generation Education* (pp. 130–151). Hershey, PA: IGI Global. doi:10.4018/978-1-5225-0483-2.ch008

Han, S. J., Lim, D. H., & Jung, E. (2019). A Collaborative Active Learning Model as a Vehicle for Online Team Learning in Higher Education. In J. Keengwe (Ed.), *Handbook of Research on Blended Learning Pedagogies and Professional Development in Higher Education* (pp. 40–59). Hershey, PA: IGI Global. doi:10.4018/978-1-5225-5557-5.ch003

Harkness, N. N. (2019). Technology in Education: The History. In D. Williams & N. Harkness (Eds.), *Diverse Learning Opportunities Through Technology-Based Curriculum Design* (pp. 1–15). Hershey, PA: IGI Global. doi:10.4018/978-1-5225-5519-3.ch001

Havard, B., & Podsiad, M. (2017). Wearable Computers. In T. Kidd & L. Morris Jr., (Eds.), *Handbook of Research on Instructional Systems and Educational Technology* (pp. 356–365). Hershey, PA: IGI Global. doi:10.4018/978-1-5225-2399-4.ch030

Hendrix, T. J. (2019). Unconventional Delivery: Developing and Implementing Service-Learning in an Online Course. In J. Keengwe (Ed.), *Handbook of Research on Blended Learning Pedagogies and Professional Development in Higher Education* (pp. 259–273). Hershey, PA: IGI Global. doi:10.4018/978-1-5225-5557-5.ch014

Hendryx, J. D. (2017). The Technology Segment of a Methods Course: Its Impact on Teaching Realities and Imagined Needs. In C. Lin, D. Zhang, & B. Zheng (Eds.), *Preparing Foreign Language Teachers for Next-Generation Education* (pp. 226–244). Hershey, PA: IGI Global. doi:10.4018/978-1-5225-0483-2.ch012

Homp, M. (2019). Facilitating Active Learning and Collaboration in Online Mathematics Content Courses for Secondary Teachers. In J. Keengwe (Ed.), *Handbook of Research on Blended Learning Pedagogies and Professional Development in Higher Education* (pp. 139–155). Hershey, PA: IGI Global. doi:10.4018/978-1-5225-5557-5.ch008

Hricko, M. (2017). Personal Learning Environments. In T. Kidd & L. Morris Jr., (Eds.), *Handbook of Research on Instructional Systems and Educational Technology* (pp. 236–248). Hershey, PA: IGI Global. doi:10.4018/978-1-5225-2399-4.ch021

Jantjies, M., & Joy, M. (2017). Teaching Through Mobile Technology: A Reflection From High School Studies in South Africa. In T. Kidd & L. Morris Jr., (Eds.), *Handbook of Research on Instructional Systems and Educational Technology* (pp. 299–312). Hershey, PA: IGI Global. doi:10.4018/978-1-5225-2399-4.ch026

Johnson, A. M., Tighe, E. L., Jacovina, M. E., Jackson, G. T., & McNamara, D. S. (2018). Development of an Interactive Narrative for Adult Literacy Learners. In R. Roscoe, S. Craig, & I. Douglas (Eds.), *End-User Considerations in Educational Technology Design* (pp. 238–262). Hershey, PA: IGI Global. doi:10.4018/978-1-5225-2639-1.ch012

Jones, C. M. (2017). Implementation of Online Education for K-12 School Children. In M. Grassetti & S. Brookby (Eds.), *Advancing Next-Generation Teacher Education through Digital Tools and Applications* (pp. 200–217). Hershey, PA: IGI Global. doi:10.4018/978-1-5225-0965-3.ch011

Kalaian, S. A. (2017). Distance and Online Learning. In T. Kidd & L. Morris Jr., (Eds.), *Handbook of Research on Instructional Systems and Educational Technology* (pp. 23–36). Hershey, PA: IGI Global. doi:10.4018/978-1-5225-2399-4.ch003

Kao, Y. S., Matlen, B. J., Tiu, M., & Li, L. (2018). Logic Models as a Framework for Iterative User Research in Educational Technology: Illustrative Cases. In R. Roscoe, S. Craig, & I. Douglas (Eds.), *End-User Considerations in Educational Technology Design* (pp. 52–75). Hershey, PA: IGI Global. doi:10.4018/978-1-5225-2639-1.ch003

Kazakoff, E. R., Orkin, M., Bundschuh, K., & Schechter, R. L. (2018). Fostering Engagement in Educational Technologies Through Developmental Theory and Program Data. In R. Roscoe, S. Craig, & I. Douglas (Eds.), *End-User Considerations in Educational Technology Design* (pp. 99–122). Hershey, PA: IGI Global. doi:10.4018/978-1-5225-2639-1.ch005

Kenttälä, V., Rousi, R., & Kankaanranta, M. (2017). Towards the Learning Experience Technology Usability Framework. In T. Kidd & L. Morris Jr., (Eds.), *Handbook of Research on Instructional Systems and Educational Technology* (pp. 128–140). Hershey, PA: IGI Global. doi:10.4018/978-1-5225-2399-4.ch012

Khan, M. S., Shamim, M. R., & Nambobi, M. (2018). Learning Styles and Online Tools: How to Construct an Effective Online Learning Environment. In A. Kumar (Ed.), *Optimizing Student Engagement in Online Learning Environments* (pp. 147–162). Hershey, PA: IGI Global. doi:10.4018/978-1-5225-3634-5.ch007

Khan, M. Z., & Alginahi, Y. M. (2018). New Frontiers for E-Learning in Education: A Big Data Application. In A. Kumar (Ed.), *Optimizing Student Engagement in Online Learning Environments* (pp. 220–240). Hershey, PA: IGI Global. doi:10.4018/978-1-5225-3634-5.ch010

Kidd, T. T. (2017). The Rhetoric of Fear: Voices and Stories Told of Faculty Who Engage in Online Teaching. In T. Kidd & L. Morris Jr., (Eds.), *Handbook of Research on Instructional Systems and Educational Technology* (pp. 69–77). Hershey, PA: IGI Global. doi:10.4018/978-1-5225-2399-4.ch007

Kidd, T. T., & Hodges, J. R. (2017). Social Media, Mobile Technology, and New Learning Opportunities: Implications for Social Justice and Educational Spaces in Schools. In T. Kidd & L. Morris Jr., (Eds.), *Handbook of Research on Instructional Systems and Educational Technology* (pp. 274–290). Hershey, PA: IGI Global. doi:10.4018/978-1-5225-2399-4.ch024

Koç, S., & Boboc, M. (2017). Structuring Online Instruction by Dynamic Design, Delivery, and Assessment. In T. Kidd & L. Morris Jr., (Eds.), *Handbook of Research on Instructional Systems and Educational Technology* (pp. 78–88). Hershey, PA: IGI Global. doi:10.4018/978-1-5225-2399-4.ch008

Kowalsky, M., & Long, L. (2019). Teacher Education Discussion Module: Online Teacher Chats About Technology in Schools. In D. Williams & N. Harkness (Eds.), *Diverse Learning Opportunities Through Technology-Based Curriculum Design* (pp. 133–146). Hershey, PA: IGI Global. doi:10.4018/978-1-5225-5519-3.ch006

Kula, I., Branaghan, R. J., Atkinson, R. K., & Roscoe, R. D. (2018). Assessing User Experience via Biometric Sensor Affect Detection. In R. Roscoe, S. Craig, & I. Douglas (Eds.), *End-User Considerations in Educational Technology Design* (pp. 123–139). Hershey, PA: IGI Global. doi:10.4018/978-1-5225-2639-1.ch006

Kumar, A. V., & Sundar, P. V. (2018). Impact of Student Engagement in Online Learning Environments. In A. Kumar (Ed.), *Optimizing Student Engagement in Online Learning Environments* (pp. 1–27). Hershey, PA: IGI Global. doi:10.4018/978-1-5225-3634-5.ch001

Lawrence, J. E. (2018). Factors Influencing Teachers' Integration of ICT in Teaching and Learning. *International Journal of Adult Vocational Education and Technology*, *9*(2), 48–63. doi:10.4018/IJAVET.2018040104

Lee, K. W., & James, C. C. (2018). Exploring a Transformative Teacher Professional Development Model to Engender Technology Integration in the 21st Century ESL Language Classrooms. *International Journal of Computer-Assisted Language Learning and Teaching*, *8*(4), 13–31. doi:10.4018/IJCALLT.2018100102

Lewis, T. D., Burks, B. A., Thompson, C. N., & Austin, S. M. (2019). Technology Impact on K-12 Education. In D. Williams & N. Harkness (Eds.), *Diverse Learning Opportunities Through Technology-Based Curriculum Design* (pp. 69–95). Hershey, PA: IGI Global. doi:10.4018/978-1-5225-5519-3.ch004

Li, J. (2017). Social Media and Foreign Language Teacher Education: Beliefs and Practices. In C. Lin, D. Zhang, & B. Zheng (Eds.), *Preparing Foreign Language Teachers for Next-Generation Education* (pp. 261–277). Hershey, PA: IGI Global. doi:10.4018/978-1-5225-0483-2.ch014

Lindsay, W. K. (2017). A Summary of Four Key Issues Affecting Distance Education. In T. Kidd & L. Morris Jr., (Eds.), *Handbook of Research on Instructional Systems and Educational Technology* (pp. 37–44). Hershey, PA: IGI Global. doi:10.4018/978-1-5225-2399-4.ch004

Little, J. M., & Fox, P. S. (2018). Developing an Engaging Online Engineering Ethics Course for Future Engineers. In A. Kumar (Ed.), *Optimizing Student Engagement in Online Learning Environments* (pp. 241–265). Hershey, PA: IGI Global. doi:10.4018/978-1-5225-3634-5.ch011

Liu, H., Lin, C., Zhang, D., & Zheng, B. (2017). Language Teachers' Perceptions of External and Internal Factors in Their Instructional (Non-) Use of Technology. In C. Lin, D. Zhang, & B. Zheng (Eds.), *Preparing Foreign Language Teachers for Next-Generation Education* (pp. 56–73). Hershey, PA: IGI Global. doi:10.4018/978-1-5225-0483-2.ch004

Liu, M. (2017). Mentoring Preservice EFL Teachers for Technology Integration: A Cloud-Based Internship Project. In C. Lin, D. Zhang, & B. Zheng (Eds.), *Preparing Foreign Language Teachers for Next-Generation Education* (pp. 175–197). Hershey, PA: IGI Global. doi:10.4018/978-1-5225-0483-2.ch010

Liu, Y. (2017). Gender Difference in Perception and Use of Social Media Tools. In T. Kidd & L. Morris Jr., (Eds.), *Handbook of Research on Instructional Systems and Educational Technology* (pp. 249–262). Hershey, PA: IGI Global. doi:10.4018/978-1-5225-2399-4.ch022

Lock, J. V., Johnson, C., Altowairiki, N., Burns, A., Hill, L., & Ostrowski, C. P. (2019). Enhancing Instructor Capacity Through the Redesign of Online Practicum Course Environments Using Universal Design for Learning. In J. Keengwe (Ed.), *Handbook of Research on Blended Learning Pedagogies and Professional Development in Higher Education* (pp. 1–20). Hershey, PA: IGI Global. doi:10.4018/978-1-5225-5557-5.ch001

López, C. M., & Hernández, P. G. (2017). Teacher, Students, and MOOCs: Innovating and Researching Teacher Training. In M. Grassetti & S. Brookby (Eds.), *Advancing Next-Generation Teacher Education through Digital Tools and Applications* (pp. 218–243). Hershey, PA: IGI Global. doi:10.4018/978-1-5225-0965-3.ch012

Mac Mahon, B., Grádaigh, S. Ó., & Ghuidhir, S. N. (2018). From iTE to NQT: Evaluating Newly Qualified Teachers' Use of Mobile Technology in Their First Two Years of Teaching. *International Journal of Mobile and Blended Learning, 10*(2), 8–19. doi:10.4018/IJMBL.2018040102

Maher, D. (2019). The Use of Course Management Systems in Pre-Service Teacher Education. In J. Keengwe (Ed.), *Handbook of Research on Blended Learning Pedagogies and Professional Development in Higher Education* (pp. 196–213). Hershey, PA: IGI Global. doi:10.4018/978-1-5225-5557-5.ch011

Manzoor, A. (2017). Blended Learning and Distance Education: Enablers and Barriers. In T. Kidd & L. Morris Jr., (Eds.), *Handbook of Research on Instructional Systems and Educational Technology* (pp. 45–57). Hershey, PA: IGI Global. doi:10.4018/978-1-5225-2399-4.ch005

Manzoor, A. (2017). Blended Learning in Higher Education: A Developing Country Perspective. In T. Kidd & L. Morris Jr., (Eds.), *Handbook of Research on Instructional Systems and Educational Technology* (pp. 58–68). Hershey, PA: IGI Global. doi:10.4018/978-1-5225-2399-4.ch006

Markman, K. M., Ferrarini, M., & Deschenes, A. H. (2018). User Testing and Iterative Design in the Academic Library: A Case Study. In R. Roscoe, S. Craig, & I. Douglas (Eds.), *End-User Considerations in Educational Technology Design* (pp. 160–183). Hershey, PA: IGI Global. doi:10.4018/978-1-5225-2639-1.ch008

Mavromoustakos, S., & Kamal, A. (2018). Student Engagement Practices for Computer Science Students in Online Learning Environments. In A. Kumar (Ed.), *Optimizing Student Engagement in Online Learning Environments* (pp. 83–99). Hershey, PA: IGI Global. doi:10.4018/978-1-5225-3634-5.ch004

McCarthy, J. (2018). Blended Learning Strategies for Engaging Diverse Student Cohorts in Higher Education. *International Journal of Teacher Education and Professional Development, 1*(2), 29–45. doi:10.4018/IJTEPD.2018070103

Militello, M., Tredway, L., & Jones, K. D. (2019). A Reimagined EdD: Participatory, Progressive Online Pedagogy. In J. Keengwe (Ed.), *Handbook of Research on Blended Learning Pedagogies and Professional Development in Higher Education* (pp. 214–243). Hershey, PA: IGI Global. doi:10.4018/978-1-5225-5557-5.ch012

Mirabolghasemi, M., Iahad, N. A., & Choshaly, S. H. (2018). Cloud Computing Assessment for Students' Social Presence in Relation to Satisfaction and Perceived Learning. In A. Kumar (Ed.), *Optimizing Student Engagement in Online Learning Environments* (pp. 59–82). Hershey, PA: IGI Global. doi:10.4018/978-1-5225-3634-5.ch003

Morris, L. R., Morse, C., & Jones, T. K. (2017). How Virtual Work Informs Virtual Learning. In T. Kidd & L. Morris Jr., (Eds.), *Handbook of Research on Instructional Systems and Educational Technology* (pp. 216–227). Hershey, PA: IGI Global. doi:10.4018/978-1-5225-2399-4.ch019

Mugayitoglu, B., & Kush, J. C. (2017). The Role of Computational Thinking in the Preparation of Pre-Service Teachers. In T. Kidd & L. Morris Jr., (Eds.), *Handbook of Research on Instructional Systems and Educational Technology* (pp. 340–355). Hershey, PA: IGI Global. doi:10.4018/978-1-5225-2399-4.ch029

Mutlu-Bayraktar, D., & Yilmaz, O. (2017). Impact of Kinect Exergame on Mental Computation Speed and Achievement. In T. Kidd & L. Morris Jr., (Eds.), *Handbook of Research on Instructional Systems and Educational Technology* (pp. 186–201). Hershey, PA: IGI Global. doi:10.4018/978-1-5225-2399-4.ch017

Nakayama, M., Mutsuura, K., & Yamamoto, H. (2017). Effectiveness of Student's Note-Taking Activities and Characteristics of Their Learning Performance in Two Types of Online Learning. *International Journal of Distance Education Technologies, 15*(3), 47–64. doi:10.4018/IJDET.2017070104

Naylor, A., & Gibbs, J. (2018). Deep Learning: Enriching Teacher Training through Mobile Technology and International Collaboration. *International Journal of Mobile and Blended Learning, 10*(1), 62–77. doi:10.4018/IJMBL.2018010105

Ngigi, S. K., & Obura, E. A. (2019). Blended Learning in Higher Education: Challenges and Opportunities. In J. Keengwe (Ed.), *Handbook of Research on Blended Learning Pedagogies and Professional Development in Higher Education* (pp. 290–306). Hershey, PA: IGI Global. doi:10.4018/978-1-5225-5557-5.ch016

Niblett, S. L., & Rakes, M. L. (2017). Social Media and Technology May Change the Culture of Rape on College Campuses. In T. Kidd & L. Morris Jr., (Eds.), *Handbook of Research on Instructional Systems and Educational Technology* (pp. 263–273). Hershey, PA: IGI Global. doi:10.4018/978-1-5225-2399-4.ch023

Noble, A., McQuillan, P., Wangia, S., & Soules, K. (2017). The Emergence of Student-Centered Teaching in Professional Learning Networks on Twitter: The Role of Choice and Voice. In M. Grassetti & S. Brookby (Eds.), *Advancing Next-Generation Teacher Education through Digital Tools and Applications* (pp. 179–199). Hershey, PA: IGI Global. doi:10.4018/978-1-5225-0965-3.ch010

Nordick, E. A. (2019). Media Literacy Instruction Using Technology. In D. Williams & N. Harkness (Eds.), *Diverse Learning Opportunities Through Technology-Based Curriculum Design* (pp. 217–237). Hershey, PA: IGI Global. doi:10.4018/978-1-5225-5519-3.ch010

Obel-Omia, C. H. (2017). Developing Digital Literacy: Preparing Next-Generation Elementary Education Literacy Teachers. In M. Grassetti & S. Brookby (Eds.), *Advancing Next-Generation Teacher Education through Digital Tools and Applications* (pp. 103–118). Hershey, PA: IGI Global. doi:10.4018/978-1-5225-0965-3.ch006

Ouyang, F., & Scharber, C. (2018). Adapting the TPACK Framework for Online Teaching Within Higher Education. *International Journal of Online Pedagogy and Course Design*, *8*(1), 42–59. doi:10.4018/IJOPCD.2018010104

Palahicky, S., DesBiens, D., Jeffery, K., & Webster, K. S. (2019). Pedagogical Values in Online and Blended Learning Environments in Higher Education. In J. Keengwe (Ed.), *Handbook of Research on Blended Learning Pedagogies and Professional Development in Higher Education* (pp. 79–101). Hershey, PA: IGI Global. doi:10.4018/978-1-5225-5557-5.ch005

Palilonis, J. A. (2017). Multimedia Active Reading: A Framework for Understanding Learning With Tablet Textbooks. In T. Kidd & L. Morris Jr., (Eds.), *Handbook of Research on Instructional Systems and Educational Technology* (pp. 153–163). Hershey, PA: IGI Global. doi:10.4018/978-1-5225-2399-4.ch014

Paracha, S., Inuoue, A., & Jehanzeb, S. (2018). Detecting Online Learners' Reading Ability via Eye-Tracking. In A. Kumar (Ed.), *Optimizing Student Engagement in Online Learning Environments* (pp. 163–185). Hershey, PA: IGI Global. doi:10.4018/978-1-5225-3634-5.ch008

Paracha, S., Takahara, T., & Jehanzeb, S. (2018). Cultural Implications for Student Engagement in Online Learning. In A. Kumar (Ed.), *Optimizing Student Engagement in Online Learning Environments* (pp. 28–58). Hershey, PA: IGI Global. doi:10.4018/978-1-5225-3634-5.ch002

Parlakkılıç, A. (2017). Modular E-Learning Course Design. In T. Kidd & L. Morris Jr., (Eds.), *Handbook of Research on Instructional Systems and Educational Technology* (pp. 228–235). Hershey, PA: IGI Global. doi:10.4018/978-1-5225-2399-4.ch020

Paynter, K. (2017). Elementary Library Media Specialists' Roles in the Implementation of the Common Core State Standards. In M. Grassetti & S. Brookby (Eds.), *Advancing Next-Generation Teacher Education through Digital Tools and Applications* (pp. 262–283). Hershey, PA: IGI Global. doi:10.4018/978-1-5225-0965-3.ch014

Penland, J. L., Laviers, K., Bassham, E., & Nnochiri, V. (2019). Virtual Learning: A Study of Virtual Reality for Distance Education. In J. Keengwe (Ed.), *Handbook of Research on Blended Learning Pedagogies and Professional Development in Higher Education* (pp. 156–176). Hershey, PA: IGI Global. doi:10.4018/978-1-5225-5557-5.ch009

Perez, G. M. (2017). Mentoring Teacher Assistants to Use Online Tools. In C. Lin, D. Zhang, & B. Zheng (Eds.), *Preparing Foreign Language Teachers for Next-Generation Education* (pp. 245–260). Hershey, PA: IGI Global. doi:10.4018/978-1-5225-0483-2.ch013

Pima, J. M., Odetayo, M., Iqbal, R., & Sedoyeka, E. (2018). A Thematic Review of Blended Learning in Higher Education. *International Journal of Mobile and Blended Learning, 10*(1), 1–11. doi:10.4018/IJMBL.2018010101

Politis, D., Stagiopoulos, P., Aidona, S., Kyriafinis, G., & Constantinidis, I. (2018). Autonomous Learning and Skill Accreditation: A Paradigm for Medical Studies. In A. Kumar (Ed.), *Optimizing Student Engagement in Online Learning Environments* (pp. 266–296). Hershey, PA: IGI Global. doi:10.4018/978-1-5225-3634-5.ch012

Polly, D. (2019). Leveraging Elements of Understanding by Design to Develop Elementary Education Candidates' TPACK in Mathematics. In D. Williams & N. Harkness (Eds.), *Diverse Learning Opportunities Through Technology-Based Curriculum Design* (pp. 147–162). Hershey, PA: IGI Global. doi:10.4018/978-1-5225-5519-3.ch007

Prion, S., & Mitchell, M. (2019). Content Considerations for Blended Learning Experiences. In J. Keengwe (Ed.), *Handbook of Research on Blended Learning Pedagogies and Professional Development in Higher Education* (pp. 102–121). Hershey, PA: IGI Global. doi:10.4018/978-1-5225-5557-5.ch006

Rahimi, M., & Pourshahbaz, S. (2019). *English as a Foreign Language Teachers' TPACK: Emerging Research and Opportunities* (pp. 1–192). Hershey, PA: IGI Global. doi:10.4018/978-1-5225-6267-2

Rangara-Omol, T. (2019). Understanding Student Support: The Link Between Faculty, Student, and Online Learning. In J. Keengwe (Ed.), *Handbook of Research on Blended Learning Pedagogies and Professional Development in Higher Education* (pp. 307–324). Hershey, PA: IGI Global. doi:10.4018/978-1-5225-5557-5.ch017

Riel, J., Lawless, K. A., & Brown, S. W. (2017). Defining and Designing Responsive Online Professional Development (ROPD): A Framework to Support Curriculum Implementation. In T. Kidd & L. Morris Jr., (Eds.), *Handbook of Research on Instructional Systems and Educational Technology* (pp. 104–115). Hershey, PA: IGI Global. doi:10.4018/978-1-5225-2399-4.ch010

Ritter, S., Murray, R. C., & Hausmann, R. G. (2018). Educational Software Design: Education, Engagement, and Productivity Concerns. In R. Roscoe, S. Craig, & I. Douglas (Eds.), *End-User Considerations in Educational Technology Design* (pp. 35–51). Hershey, PA: IGI Global. doi:10.4018/978-1-5225-2639-1.ch002

Roberts, J., & Kidd, T. T. (2017). Millennials, Digital Natives, and the Emergence of New Educational Spaces. In T. Kidd & L. Morris Jr., (Eds.), *Handbook of Research on Instructional Systems and Educational Technology* (pp. 1–10). Hershey, PA: IGI Global. doi:10.4018/978-1-5225-2399-4.ch001

Roscoe, R. D., Branaghan, R. J., Cooke, N. J., & Craig, S. D. (2018). Human Systems Engineering and Educational Technology. In R. Roscoe, S. Craig, & I. Douglas (Eds.), *End-User Considerations in Educational Technology Design* (pp. 1–34). Hershey, PA: IGI Global. doi:10.4018/978-1-5225-2639-1.ch001

Rotich, J. P., & Elliott, G. (2019). Teaching First Aid, CPR, and AED Using Blended Learning in Academic Settings. In J. Keengwe (Ed.), *Handbook of Research on Blended Learning Pedagogies and Professional Development in Higher Education* (pp. 325–339). Hershey, PA: IGI Global. doi:10.4018/978-1-5225-5557-5.ch018

Rusko, R. (2017). Interactive Online Learning: Solution to Learn Entrepreneurship in Sparsely Populated Areas? *International Journal of Online Pedagogy and Course Design*, *7*(3), 54–69. doi:10.4018/IJOPCD.2017070104

Santos, A. (2017). Instructional Strategies for Game-Based Learning. In T. Kidd & L. Morris Jr., (Eds.), *Handbook of Research on Instructional Systems and Educational Technology* (pp. 164–173). Hershey, PA: IGI Global. doi:10.4018/978-1-5225-2399-4.ch015

Saravanan, K. (2018). MOOC for Student Learning and Active Engagement. In A. Kumar (Ed.), *Optimizing Student Engagement in Online Learning Environments* (pp. 127–146). Hershey, PA: IGI Global. doi:10.4018/978-1-5225-3634-5.ch006

Schroeder, N. L., Oliver, A., Deffet, K., & Morgan, J. (2018). Development of an Educational Video Game That Can Be Modified by End-Users. In R. Roscoe, S. Craig, & I. Douglas (Eds.), *End-User Considerations in Educational Technology Design* (pp. 184–204). Hershey, PA: IGI Global. doi:10.4018/978-1-5225-2639-1.ch009

Schugar, J., & Schugar, H. (2018). eTexts and Teacher Education: Considerations for Text Structure and Purpose in Mobile Pedagogy. *International Journal of Mobile and Blended Learning*, *10*(2), 78–87. doi:10.4018/IJMBL.2018040107

Shambaugh, N. (2019). Applying a Teaching Decision Cycle to the Design of Online Learning Within Faculty Professional Development. In J. Keengwe (Ed.), *Handbook of Research on Blended Learning Pedagogies and Professional Development in Higher Education* (pp. 21–39). Hershey, PA: IGI Global. doi:10.4018/978-1-5225-5557-5.ch002

Stone, M. L., Kent, K. M., Roscoe, R. D., Corley, K. M., Allen, L. K., & McNamara, D. S. (2018). The Design Implementation Framework: Iterative Design From the Lab to the Classroom. In R. Roscoe, S. Craig, & I. Douglas (Eds.), *End-User Considerations in Educational Technology Design* (pp. 76–98). Hershey, PA: IGI Global. doi:10.4018/978-1-5225-2639-1.ch004

Stover, S., Heilmann, S. G., & Hubbard, A. R. (2018). Student Perceptions Regarding Clickers: The Efficacy of Clicker Technologies. In R. Roscoe, S. Craig, & I. Douglas (Eds.), *End-User Considerations in Educational Technology Design* (pp. 291–315). Hershey, PA: IGI Global. doi:10.4018/978-1-5225-2639-1.ch014

Sun, Z., You, J., Song, W., Qu, Z., & Luo, L. (2017). Identifying the Contributors to Improve Mobile-Based TPACK Competency of Elementary School Teachers in China. In C. Lin, D. Zhang, & B. Zheng (Eds.), *Preparing Foreign Language Teachers for Next-Generation Education* (pp. 74–91). Hershey, PA: IGI Global. doi:10.4018/978-1-5225-0483-2.ch005

Taylor, M., Atas, S., & Ghani, S. (2017). Exploring the Experiences of Students and Professors in a Blended Learning Graduate Program: A Case Study of a Faculty of Education. *International Journal of Mobile and Blended Learning*, *9*(1), 1–15. doi:10.4018/IJMBL.2017010101

Taylor, M., Ghani, S., Atas, S., & Fairbrother, M. (2018). A Pathway Towards Implementation of Blended Learning in a Medium Sized Canadian University. *International Journal of Online Pedagogy and Course Design*, *8*(1), 60–76. doi:10.4018/IJOPCD.2018010105

Taylor, M., Vaughan, N., Ghani, S. K., Atas, S., & Fairbrother, M. (2018). Looking Back and Looking Forward: A Glimpse of Blended Learning in Higher Education From 2007-2017. *International Journal of Adult Vocational Education and Technology*, *9*(1), 1–14. doi:10.4018/IJAVET.2018010101

Taylor, M. C., Atas, S., & Ghani, S. (2019). Alternate Dimensions of Cognitive Presence for Blended Learning in Higher Education. *International Journal of Mobile and Blended Learning*, *11*(2), 1–18. doi:10.4018/IJMBL.2019040101

Teng, Y. (2017). EFL Teachers' Knowledge of Technology in China: Issues and Challenges. In C. Lin, D. Zhang, & B. Zheng (Eds.), *Preparing Foreign Language Teachers for Next-Generation Education* (pp. 23–37). Hershey, PA: IGI Global. doi:10.4018/978-1-5225-0483-2.ch002

Tietjen, P., & Akgun, M. (2018). Deciphering Conflicts of Affordances Through a Design-Based Approach. In R. Roscoe, S. Craig, & I. Douglas (Eds.), *End-User Considerations in Educational Technology Design* (pp. 263–290). Hershey, PA: IGI Global. doi:10.4018/978-1-5225-2639-1.ch013

Torres-Coronas, T., & Vidal-Blasco, M. (2017). MOOC and Blended Learning Models: Analysis from a Stakeholders' Perspective. *International Journal of Information and Communication Technology Education, 13*(3), 88–99. doi:10.4018/IJICTE.2017070107

Trust, T., Maloy, R. W., & Kommers, S. (2017). From 2D Thinking to 3D Printing: Preservice and In-Service Teacher Teams Explore a New Technology. In M. Grassetti & S. Brookby (Eds.), *Advancing Next-Generation Teacher Education through Digital Tools and Applications* (pp. 161–178). Hershey, PA: IGI Global. doi:10.4018/978-1-5225-0965-3.ch009

Tsai, T. P., Lin, L. C., & Lin, J. (2019). A Study on the Preview Effectiveness of Learning Contents in ePUB3 eBook-Based Flip Blended Learning Models. *International Journal of Mobile and Blended Learning, 11*(2), 50–67. doi:10.4018/IJMBL.2019040104

Vishkaie, R., Shively, K., & Wessel Powell, C. (2018). Perceptions of Digital Tools and Creativity in the Classroom. *International Journal of Digital Literacy and Digital Competence, 9*(4), 1–18. doi:10.4018/IJDLDC.2018100101

Walker, E., Wylie, R., Danielescu, A., Rodriguez, J. P. III, & Finn, E. (2018). Balancing Student Needs and Learning Theory in a Social Interactive Postdigital Textbook. In R. Roscoe, S. Craig, & I. Douglas (Eds.), *End-User Considerations in Educational Technology Design* (pp. 141–159). Hershey, PA: IGI Global. doi:10.4018/978-1-5225-2639-1.ch007

Wang, W., & Feng, L. (2017). Technology Standards for Chinese Language Teacher Education. In C. Lin, D. Zhang, & B. Zheng (Eds.), *Preparing Foreign Language Teachers for Next-Generation Education* (pp. 38–54). Hershey, PA: IGI Global. doi:10.4018/978-1-5225-0483-2.ch003

Ward-Jackson, J., & Yu, C. (2019). Impact of Online Learning in K-12: Effectiveness, Challenges, and Limitations for Online Instruction. In J. Keengwe (Ed.), *Handbook of Research on Blended Learning Pedagogies and Professional Development in Higher Education* (pp. 357–375). Hershey, PA: IGI Global. doi:10.4018/978-1-5225-5557-5.ch020

Webster, K. S. (2017). RSS and Syndication for Educators. In T. Kidd & L. Morris Jr., (Eds.), *Handbook of Research on Instructional Systems and Educational Technology* (pp. 366–379). Hershey, PA: IGI Global. doi:10.4018/978-1-5225-2399-4.ch031

Wilson, C., Bennett, W. Jr, Guarino, S., Bove, K., & Cain, T. L. (2018). Applying a User-Centered Design Approach to Developing Game-Based Training Environments for Aircraft Maintainers. In R. Roscoe, S. Craig, & I. Douglas (Eds.), *End-User Considerations in Educational Technology Design* (pp. 217–237). Hershey, PA: IGI Global. doi:10.4018/978-1-5225-2639-1.ch011

Wilson, N. (2017). Sociotechnical and Pedagogical Barriers to Technology Integration. In M. Grassetti & S. Brookby (Eds.), *Advancing Next-Generation Teacher Education through Digital Tools and Applications* (pp. 19–37). Hershey, PA: IGI Global. doi:10.4018/978-1-5225-0965-3.ch002

Yonemoto, K., Tsuda, A., & Hayashi, H. (2017). Implementing a Flipped Classroom in Teaching Second Language Pronunciation: Challenges, Solutions, and Expectations. In C. Lin, D. Zhang, & B. Zheng (Eds.), *Preparing Foreign Language Teachers for Next-Generation Education* (pp. 114–129). Hershey, PA: IGI Global. doi:10.4018/978-1-5225-0483-2.ch007

Yu, C., & Franz, D. P. (2017). Visiting Technological Pedagogical and Content Knowledge (TPACK): Issues and Challenges for Teachers' Professional Development. In T. Kidd & L. Morris Jr., (Eds.), *Handbook of Research on Instructional Systems and Educational Technology* (pp. 380–391). Hershey, PA: IGI Global. doi:10.4018/978-1-5225-2399-4.ch032

Zhang, Y., & Deroo, M. (2017) Technology Instruction in Language Teacher Education Programs. In C. Lin, D. Zhang, & B. Zheng (Eds.), *Preparing Foreign Language Teachers for Next-Generation Education* (pp. 1–22). Hershey, PA: IGI Global. doi:10.4018/978-1-5225-0483-2.ch001

Ziliak, E. M. (2019). Calculus 1 Course Comparison: Online/Blended or Flipped? In J. Keengwe (Ed.), *Handbook of Research on Blended Learning Pedagogies and Professional Development in Higher Education* (pp. 244–258). Hershey, PA: IGI Global. doi:10.4018/978-1-5225-5557-5.ch013

About the Author

Manjit Singh Sidhu is currently a professor attached to the College of Computing & Informatics, Department of Graphics and Multimedia Department, University Tenaga Nasional, Malaysia. He received his BSc. (Hons) degree in Computer Science from the University of Wolverhampton, UK in 1997, Masters in Information Technology from University Putra Malaysia in 2000 and PhD in Multimedia from the University of Malaya in 2008. His background covers a wide range of computer science and educational experiences, initially teaching undergraduates and supervising postgraduates both Masters and PhDs. He has written and published several conference proceedings, journals, book chapters both locally and internationally and a full book with IGI Global. He has also won a number of research innovation and invention awards both locally and internationally and filed for patterns for his work. He is a Chartered IT Professional with the British Computer Society (BCS) (UK), Fellow of BCS, a senior member of Institute of Electrical and Electronics Engineers (IEEE), Computer and Communications Society, Malaysian National Computer Confederation (MNCC), Fellow of the Malaysian Scientific Association (MSA) and registered Professional Technologists with the Malaysian Board of Technologists (MBOT) Malaysia. His research focuses on patterns of interactions for multimedia, virtual and augmented reality applications, visualization and computer animations.

Index

O

P

Q

S

T

U

V

W

www.ingramcontent.com/pod-product-compliance
Lightning Source LLC
LaVergne TN
LVHW081327300926
849161LV00008B/1904